高等职业教育工业机器人技术专业系列教材

工业机器人专业英语

吕亚男　温贻芳　编

机械工业出版社

本书共 22 章，内容包括 Robot、Types of Robot、Industrial Robot、Application of Industrial Robot、Industrial Robot Manufacturer、Manipulator of Industrial Robot、Technical Parameter、Robot Coordinates、Drive system、Transmission Component、Sensors、Robot-Computer Interface、Program and Programming、Machine Visions、Programmable Logic Controller、Robotic Technician and Engineer、Robotic Research and Development、Robot Manuals、Safety Considerations、Robot Maintenance、Industrial Robot and Industry 4.0 和 Robot Policy。每章包括课文、词汇、课文重点问题和译文。

本书可作为高等职业院校工业机器人技术、机电一体化技术和电气自动化技术等智能制造相关专业的教材，也可供应用型本科院校工业机器人相关专业的师生以及工程技术人员学习参考。

图书在版编目（CIP）数据

工业机器人专业英语/吕亚男，温贻芳编. —北京：机械工业出版社，2021.4（2024.8 重印）
高等职业教育工业机器人技术专业系列教材
ISBN 978-7-111-67877-9

Ⅰ.①工… Ⅱ.①吕… ②温… Ⅲ.①工业机器人-英语-高等职业教育-教材 Ⅳ.①TP242.2

中国版本图书馆 CIP 数据核字（2021）第 057163 号

机械工业出版社（北京市百万庄大街 22 号　邮政编码 100037）
策划编辑：薛　礼　责任编辑：薛　礼
责任校对：李　伟　封面设计：张　静
责任印制：常天培
北京中科印刷有限公司印刷
2024 年 8 月第 1 版第 4 次印刷
184mm×260mm · 7.5 印张 · 184 千字
标准书号：ISBN 978-7-111-67877-9
定价：29.00 元

电话服务　　　　　　　　　　网络服务
客服电话：010-88361066　机　工　官　网：www.cmpbook.com
　　　　　010-88379833　机　工　官　博：weibo.com/cmp1952
　　　　　010-68326294　金　书　网：www.golden-book.com
封底无防伪标均为盗版　机工教育服务网：www.cmpedu.com

Industrial Robot

Preface

　　伴随着工业变革以及科技进步，机器人技术自从其出现以来在全球都受到了高度关注。其中，工业机器人是目前工业领域应用最广泛的机器人，大量应用在工程机械、汽车工程和电子信息等行业。工业机器人的发展和应用，不仅有效提升了工作效率和产品质量，而且全面推动了制造业的发展。目前工业机器人产业的地位已经上升到了国家层面，作为国家十大重点发展领域，大力发展、全力推进工业机器人产业，对推动产业转型、建设制造强国具有深远意义。

　　伴随着全球一体化的浪潮，我国高等职业教育也逐步趋向国际化和现代化，尤其是工业机器人技术等发展速度较快的机电类专业对人才培养中专业英语教学的要求更为急迫。目前，我国国内严重缺乏工业机器人专业英语教材，同时高等职业院校学生严重缺乏直接阅读外文材料的能力。"工业机器人专业英语"是高等职业教育工业机器人等机电类专业的专业基础课，目的是帮助高等职业院校学生理解及掌握专业英语，包括定义概念、基本理论、用途分类以及发展前景等。本书以高等职业院校教育工业机器人技术相关专业对专业英语教材的需求为出发点，从适应高等职业院校专业英语的实际教学需要考虑，较全面地涵盖了工业机器人专业英语的相关知识。本书对工业机器人的历史、发展、结构、特点、功能、应用、维护和工业4.0等内容进行了整理和精选。本书在每一章均设置了阅读理解、词汇介绍、内容分析、重点问题和中文翻译。读者通过学习，不但可以熟悉工业机器人的常用专业词汇，也可对工业机器人理论知识有进一步的了解，能够为阅读专业文献打下基础。本书可作为工业机器人技术等相关专业的专业英语教材，也可供从事机器人相关领域工作的技术人员参考。

　　本书在编写过程中得到了编者所在院校领导的高度重视和大力支持，专业老师对本书的内容提出了宝贵意见，在此表示衷心的感谢！

　　由于编者水平有限，书中难免存在一些缺点和不妥之处，请读者批评指正。

编　者

目录
Contents

Chapter **1** Robots

Objectives

After reading this chapter, you will be able to:

1) identify and discuss the key terms used.

2) understand the definition of robot.

3) know the differences between robot and human.

4) identify the positive and negative aspects of robot.

5) answer the review questions at the end of the chapter.

Reading

What is a robot? A robot is defined as a programmable multifunctional manipulator designed to move materials, parts, tools, or specialized devices. The robot performs various tasks through programmable variable motion.

The word "Robota" was first used by Karl Capek (1890—1938) in Czech in 1921. It was used to describe the machine that performed like a human, but had no human feeling. Then it was widely known as "Robot" in English. Fig. 1-1 shows the pictures of opera "Rossum's Universal Robots".

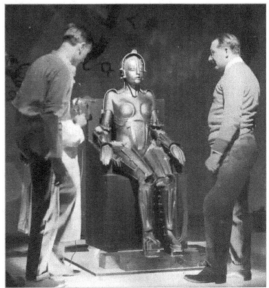

a)   b)

Fig. 1-1   Rossum's Universal Robots (R. U. R.)

A robot is designed to assist and replace human labor. It has several advantages. Firstly, robot can keep working for a long time without a break when the conditions allow. Secondly, the performance efficiency of robot is much higher than human. Thirdly, robot is flexible to control by the programming based on task requirements. Lastly, robot has low requirement for the working environment, and that is also why it is widely applied in the specific and dangerous environments.

However, the advantages are always companied with the disadvantages. Reasonable considerations should be required before using robot to replace human labor. As a kind of machine, the robot is controlled by high-skilled workers, therefore, the requirements for the technicians are higher. Furthermore, the initial installation and programming cost of the robot is high. The most important issue is the downtime due to malfuction or overhaul. The repair and maintenance of robot highly depend on the quality of manufactured products, and it may affect the production schedule of the entire plant.

Vocabulary

robot   /'rəʊbət/ *n.* 机器人；机械般工作的人

programmable   /ˌprəʊ'græməbl/ *adj.* ［计］可编程的；可设计的

multifunctional   /ˌmʌlti'fʌŋkʃənl/ *adj.* 多功能的

manipulator   /mə'nɪpjʊleɪtə(r)/ *n.* 操纵器；操作者

device   /dɪ'vaɪs/ *n.* 装置；策略；图案；设备；终端

sense   /sens/ *n.* 感觉，功能；理智 *vt.* 感觉到；检测

assist   /ə'sɪst/ *v.* 参加，出席；协助；有助益 *n.* 帮助；助攻；辅助机械装置

advantage   /əd'vɑːntɪdʒ/ *n.* 优势；有利条件 *vt.* 有利于；使处于优势 *vi.* 获利

efficiency   /ɪ'fɪʃ(ə)nsɪ/ *n.* 效率；效能；功效

flexible   /'fleksəb(ə)l/ *adj.* 灵活的；柔韧的；易弯曲的

surrounding   /sə'raʊndɪŋ/ *n.* 环境，周围的事物 *adj.* 周围的，附近的

environment   /ɪn'vaɪrənmənt/ *n.* 环境，外界

company   /'kʌmp(ə)nɪ/ *n.* 公司；陪伴，同伴；连队 *vt.* 陪伴 *vi.* 交往

disadvantage   /dɪsəd'vɑːntɪdʒ/ *n.* 缺点；不利条件；损失

adopt   /ə'dɒpt/ *vi.* 采取；过继 *vt.* 采取；接受；收养；正式通过

talent   /'tælənt/ *n.* 才能；天才；天资

initial   /ɪ'nɪʃəl/ *n.* 词首大写字母 *adj.* 最初的；字首的

downtime   /'daʊntaɪm/ *n.* 停工期；［电子］故障停机时间

maintenance   /'meɪntənəns/ *n.* 维护，维修；保持；生活费用

schedule   /'skedʒʊl/ *v.* 安排，预定 *n.* 计划（表）；时间表

Notes and analysis

Question 1：What is a robot?

Answer: _____

Question 2: Who gives the name of the robot?

Answer: _____

Question 3: List the main advantages of the robot.

Answer: _____

Question 4: List the main disadvantages of the robot.

Answer: _____

Question 5: What is the difference between robot and human?

Answer: _____

Question 6: What is the aim of developing robot?

Answer: _____

Translation

机　器　人

什么是机器人？机器人被定义为一种可编程的多功能机械手，被设计用于移动材料、部件、工具或专门的设备，机器人通过可编程的可变运动来执行各种任务。

1921年，捷克人卡尔·卡佩克首次用"Robota"这个词来形容机器，它的表现像人一样，但没有人类的感觉，后来在英语中被广泛称为"Robot"。图1-1所示为歌剧《罗萨姆万能机器人》剧照。

机器人是为辅助和代替人类劳动而设计的。它具有以下优势：首先，在条件允许的情况下，机器人可以长时间不间断地工作；其次，机器人的工作效率远远高于人类；第三，基于任务要求和编程指令，机器人比较灵活；最后，机器人对工作环境的要求较低，因此被广泛应用于特定的危险环境中。

然而，优点总是伴随着缺点。在采用机器人代替人工之前，应该经过认真合理的考虑。机器人作为一种机器，由技术工人控制其运行，因此对技术工人的要求较高。机器人最初的安装和编程成本很高。最重要的问题是由于故障或者检修导致的停工期。机器人的维修保养与产品的质量密切相关，可能会影响整个工厂的生产进度。

ROBOT
Chapter 2 Types of Robots

Objectives

After reading this chapter, you will be able to:

1) have an understanding of the flexibility of the robot.

2) know the types of robots.

3) understand the differences between the robots.

4) be capable of choosing suitable robot.

5) answer the review questions at the end of the chapter.

Reading

As we all know, the robot is composed of multiple subsystems, so it can be flexibly applied to various environments with the specific design and assembly. For instance, the robot can be used to move and stack the products in the warehouse, or help surgeons perform an operation in the hospital. The design and type of the robot are determined by actual needs. Normally, robots are classified into industrial robot, laboratory robot, medical robot, and explorer robot, etc.

1. Industrial robot

Robots used in the industry are normally called industrial robot, with grippers or other tools on their mechanical arms, as shown in Fig. 2-1. The arms are responsible for moving things to the target position under the motion programming, and the grippers are responsible for picking up the product and placing it in a new position. All the industrial robots can be programed and computerized. There are also many types of industrial robots, such as welding robot, assembly robot, painting robot and so on.

Fig. 2-1　An industrial robot

2. Laboratory robot

Different from the industrial robot, the shapes of laboratory robot are not fixed. Normally, a laboratory robot has microcomputer, multi-jointed arms, and vision functions depending on the specific requirement. A laboratory robot (Fig. 2-2) can be stationary or mobile, but it has only one purpose, which is to save much labor work for the repeatable lab programs.

3. Medical robot

Medical Robot (Fig. 2-3) is the robot that is normally found in the hospital and medical research institute to serve the medical performance and nursing service. The first known medical robot is named PUMA 560 in 1985. Medical robot has high working precision, and it works without fatigue. It can significantly improve the speed and quality in the surgical procedures. Furthermore, the medical robot can provide care and monitoring services for the patients and the elderly, which saves labor and responds quickly to the medical needs.

Fig. 2-2　A laboratory robot

Fig. 2-3　A medical robot

4. Explorer robot (Fig 2-4)

For the situations that are inaccessible or dangerous, such as war zones, wild field, outer space, and deep sea, explorer robots are used to explore, rescue, and collect data. Explorer robots have higher techical requirement than other types of robots, as their working environment is tough and harsh. In addition, the explrver robot must have extremely terrain mobility, multi-sensory system, strong communication capabilities, and control system to ensure their normal use. This is also the reason that the explorer robots are more expensive than other robots.

Fig. 2-4　An explorer robot

Vocabulary

subsystem /ˈsʌbsɪstəm/ *n.* 子系统；次要系统

flexible /ˈfleksəb(ə)l/ *adj.* 灵活的；柔韧的；易弯曲的

circumstance /ˈsɜːkəmstəns/ *n.* 环境；状况；境遇；命运 *vt.* 处于某种情况

design /dɪˈzaɪn/ *v.* 设计，构思；计划 *n.* 设计；构思；设计图样

assembly /əˈsemblɪ/ *n.* 装配；集会，集合 *n.* 汇编，编译

stack /stæk/ *n.* （整齐的）一堆；垛，堆 *v.* （使）成叠地放在

warehouse /ˈweəhaʊs/ *n.* 仓库；货栈；大商店 *vt.* 储入仓库

surgeon /ˈsɜːdʒ(ə)n/ *n.* 外科医生

determine /dɪˈtɜːmɪn/ *v.* （使）下决心 *vt.* 决定，确定；

classify /ˈklæsɪfaɪ/ *vt.* 分类；分等

explorer /ekˈsplɔːrə(r)/ *n.* 探险家；勘探者；探测器；［医］探针

gripper /ˈgrɪpə/ *n.* 夹子，钳子；抓器，抓爪

target /ˈtɑːgɪt/ *n.* 目标，指标；靶子 *v.* 把…作为目标；面向

weld /weld/ *n.* 焊接；焊接点 *vt.* 焊接；使结合；使成整体 *vi.* 焊牢

microcomputer /ˈmaɪkrə(ʊ)kɒmˌpjuːtə/ *n.* 微电脑；［计］微型计算机

stationary /ˈsteɪʃ(ə)n(ə)rɪ/ *n.* 不动的人 *adj.* 固定的；静止的；定居的

mobile /ˈməʊbaɪl/ *n.* 移动电话 *adj.* 可移动的；机动的；易变的；非固定的

repeatable /riˈpiːtəbl/ *adj.* 可重复的；可复验的

institute /ˈɪnstɪtjuːt/ *v.* 实行，建立 *n.* 机构，研究所，学会

serve /sɜːv/ *n.* 发球 *vi.* 服役，招待，侍候 *vt.* 招待，供应；为…服务

nursing /ˈnɜːsɪŋ/ *n.* 护理；看护；养育 *v.* 看护；养育（nurse 的 ing 形式）

precision /prɪˈsɪʒ(ə)n/ *n.* 精度，［数］精密度；精确 *adj.* 精密的，精确的

fatigue /fəˈtiːg/ *n.* 疲劳，疲乏；杂役 *adj.* 疲劳的 *vt.* 使疲劳 *vi.* 疲劳

speed /spiːd/ *v.* 快速运动；加速；（使）繁荣 *n.* 速度；进度；迅速

quality /ˈkwɒlətɪ/ *n.* 质量，［统计］品质；特性；才能 *adj.* 优质的；高品质的

surgical /ˈsɜːdʒɪk(ə)l/ *n.* 外科手术；外科病房 *adj.* 外科的；手术上的

procedure /prəˈsiːdʒə/ *n.* 程序，手续；步骤

guardian /ˈgɑːdɪən/ *n.* ［法］监护人，保护人；守护者 *adj.* 守护的

patient /ˈpeɪʃ(ə)nt/ *n.* 病人，患者；受动者，承受者 *adj.* 有耐心的，能容忍的

response /rɪˈspɒns/ *n.* 响应；反应；回答

inaccessible /ɪnəkˈsesɪb(ə)l/ *adj.* 难达到的；难接近的；难见到的

rescue /ˈreskjuː/ *n.* 营救，解救，援救 *v.* 营救，援救

gather /ˈgæðə/ *n.* 聚集 *vt.* 收集；使…聚集 *vi.* 聚集；皱起

tough /tʌf/ *adj.* 艰苦的，困难的 *vt.* 坚持；忍耐 *adv.* 强硬地，顽强地

harsh /hɑːʃ/ *adj.* 严厉的；严酷的

extreme /ɪkˈstriːm; ek-/ *n.* 极端；最大程度 *adj.* 极端的；极度的；偏激的

terrain /təˈreɪn/ *n.* ［地理］地形，地势；领域；地带
communication /kəˌmjuːnɪˈkeɪʃn/ *n.* 通信；交流
capability /ˌkeɪpəˈbɪləti/ *n.* 才能，能力；性能，容量

Notes and analysis

Question 1：How many types of robots are mentioned in this chapter, and what are they?

Answer：_____

Question 2：What is an industrial robot?

Answer：_____

Question 3：What are the main characteristics of laboratory robot?

Answer：_____

Question 4：Can the medical robot take the place of human now? Why?

Answer：_____

Question 5：How to choose a suitable explorer robot?

Answer：_____

Translation

机器人类型

众所周知，机器人是由多个子系统组成的，因此它可以通过特定的设计和装配，灵活地应用于各种环境。例如，机器人可以用来移动和堆放在仓库的产品或者帮助外科医生在医院进行手术。机器人的设计和类型是由实际需求决定的。机器人一般分为工业机器人、实验室机器人、医疗机器人、探测机器人等。

1. 工业机器人

工业领域使用的机器人通常被称为工业机器人，它们的机械手臂上有夹具或其他工具。机械手臂负责在运动规划下移动到目标位置，夹具负责将产品捡起并放置到新的位置。所有的工业机器人都可以通过编程和计算机控制。工业机器人的种类多种多样，如焊接机器人、装配机器人和喷漆机器人等。

2. 实验室机器人

与工业机器人不同，实验室机器人的形状是不固定的。实验室机器人一般具备微型处理器、多关节手臂，根据需要还具有视觉功能。实验室机器人可以是静止的，也可以是移动的，但它的用途只有一个，就是为可重复的实验室流程节省大量的人力劳动。

3. 医疗机器人

医疗机器人是在医院和医学研究机构中常见的服务于医疗工作和护理服务的机器人。第一个已知的医疗机器人是在 1985 年被命名为 PUMA 560 的机器人。医疗机器人工作精度高，且其工作无疲劳感。医疗机器人可以显著提高手术的速度和质量。此外，医疗机器人可以为患者和老年人提供护理和监护服务，节省劳动力，对医疗需求能做出快速反应。

4. 探测机器人

　　对于战争地区、野外、外层空间、深海等难以接近或危险的环境，可以使用探测机器人进行探测、救援和数据采集等工作。探测机器人的工作环境恶劣，因此它的技术要求比其他类型的机器人都要高。此外，探测机器人要具备极高的地形机动性、多传感器系统、强大的通信能力和控制系统，以保证它的正常使用。这也是探测机器人的价格比其他机器人更昂贵的原因。

Chapter 3 Industrial Robot

Objectives

After reading this chapter, you will be able to:
1) be familiar with the history of an industrial robot.
2) know the first industrial robot.
3) classify the application of industrial robot.
4) answer the review questions at the end of the chapter.

Reading

The first industrial robot, *Unimate*, was created by Joseph F. Engelberger (1925—2015, Fig. 3-1) in 1959, which joined the assembly line and performed the welding work. *Unimate*'s job was to grab the die-casting parts from the machine and weld it.

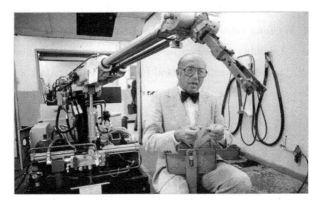

Fig. 3-1　Joseph F. Engelberger and *Unimate*

With the continuous development of the industrial robot, a variety of industrial tasks can be companied with the industrial robot. So far after more than 60 years of development, the industrial robot is reliable and easy to use.

The application of robots in industry (industrial robot) is the result of improving the quality and reducing the cost. At the beginning, the application of automated machinery is the first choice, as it can perform various industrial production operations. However, if the design of the product needs to be updated quickly, it must replace all the automated machinery, which directly leads to the much high cost.

Different from the automated machinery, industrial robot is flexible, because it can be reprogrammed according to the production requirement. Fig 3-2 shows a typical manufacturing line with industrial robots. As our discussion in Chapter 2, both advantages and disadvantages of using industrial robot should be considered before its application. There is no doubt that the industrial robot has been one of the most important parts for the modern manufacturers.

Fig. 3-2　A manufacturing line with industrial robots

The examples of the application of industrial robots are as follows:

1) Loading robot. Loading robot places products to the targeted location.

2) Palletizing robot. Palletizing robot can pack packages or move pallets.

3) Painting robot. Painting robot can paint objects like cars and furnitures.

4) Welding robot. Welding robot can weld the metallic materials.

5) Assembly robot. Assembly robot can assembly electronics and automobiles, etc.

6) Inspection robot. Inspection robot is used for the quality control with vision functions.

7) Fabricating robot. Fabricating robot can perform casting, grinding, cutting, wiring, drilling and so on.

Vocabulary

join　/dʒɔɪn/ n. 结合；连接；接合点 vi. 加入；参加；结合 vt. 参加；结合

die-casting　n. 压模法；铸造法

variety　/vəˈraɪətɪ/ n. 多样；种类；杂耍；变化，多样化

reliable　/rɪˈlaɪəb(ə)l/ n. 可靠的人 adj. 可靠的；可信赖的

application　/ˌæplɪˈkeɪʃ(ə)n/ n. 应用；申请；应用程序

manufacture　/ˌmænjuˈfæktʃə/ n. 制造；产品；制造业 vt. 制造；加工 vi. 制造

automated　/ˈɔːtəˌmeɪtɪd/ adj. 自动化的；机械化的 v. 自动操作

flexible　/ˈfleksəb(ə)l/ adj. 灵活的；柔韧的；易弯曲的

reprogram　/riːˈprəʊɡræm/ v. 为……重编程序

doubt　/daʊt/ n. 怀疑；疑问；疑惑 v. 怀疑；不信

modern　/ˈmɒd(ə)n/ n. 现代人；有思想的人 adj. 现代的，近代的；时髦的

loading　/'ləʊdɪŋ/ *n.* 装载；装货；装载的货 *v.* 装载，装填，装入

target　/'tɑːgɪt/ *v.* 面向；把……作为目标 *n.* 目标，指标；靶子

palletizing　/'pælə‚taizɪŋ/ *n.* 夹板装载；码垛堆积 *v.* 把…装在货盘上

package　/'pækɪdʒ/ *n.* 包，包裹；[计] 程序包 *vt.* 打包；将…包装

spray　/spreɪ/ *n.* 喷雾，喷雾剂；喷雾器 *vt.* 喷射 *vi.* 喷

painting　/'peɪntɪŋ/ *n.* 绘画；油画；着色 *v.* 绘画；涂色于

object　/'ɒbdʒɪkt;-dʒekt/ *n.* 目标；物体；客体 *vi.* 反对；拒绝

furniture　/'fɜːnɪtʃə/ *n.* 家具；设备；储藏物

metallic　/mə'tælɪk/ *adj.* 金属的，含金属的

material　/mə'tɪərɪəl/ *adj.* 物质的；客观存在的 *n.* 材料；用具

electronic　/ɪ‚lek'trɒnɪk/ *adj.* 电子的 *n.* 电子电路；电子器件

inspection　/ɪn'spekʃn/ *n.* 视察，检查

fabrication　/fæbrɪ'keɪʃ(ə)n/ *n.* 制造，建造；装配；伪造物

cast　/kɑːst/ *n.* 投掷，抛 *vt.* 投，抛；浇铸 *vi.* 投，计算

grind　/graɪnd/ *n.* 磨；苦工作 *vt.* 磨碎；磨快 *vi.* 磨碎；折磨

cut　/kʌt/ *v.* 割破；切下；剪切 *n.* 切，割 *adj.* 缩减的；割下的

wire　/waɪə/ *n.* 电线；金属丝；电报 *vt.* 拍电报；给…装电线 *vi.* 打电报

drill　/drɪl/ *n.* 钻子 *vt.* 钻孔；训练；条播 *vi.* 钻孔；训练

Notes and analysis

Question 1：Can industrial robots work in various circumstances? Why?

Answer：_____

Question 2：Which one is better, an automated machinery or an industrial robot? Why?

Answer：_____

Question 3：Why is it necessary to reprogram the industrial robot?

Answer：_____

Question 4：List the types of industrial robots mentioned in this chapter.

Answer：_____

Question 5：What can the fabricating robot do?

Answer：_____

Translation

工业机器人

第一台工业机器人是由约瑟夫·恩格尔伯格在 1959 年制造的，名为尤尼梅特，该机器人在生产线中进行焊接工作。尤尼梅特的工作任务是从机器上取下压铸件并对其进行焊接。

随着工业机器人的不断发展，各种工业任务中都可以见到工业机器人的参与。到目前为

止，工业机器人经过了 60 多年的发展，运行可靠且易于使用。

机器人在工业上的应用是为了提高产品质量和降低制造成本。起初，自动化生产机器应用是首选，因为它可以执行各种工业生产操作。但是，如果产品的设计需要快速更新，那么全部原有自动化生产设备都会被替换掉，直接导致生产成本过高。

与自动化生产机器不同的是，工业机器人具有较强的灵活性，它可以根据生产需要进行重新编程。图 3-2 所示为一条典型的、有工业机器人参与的生产线。针对第 2 章介绍的工业机器人的优点和缺点，应用前必须予以考虑。毫无疑问，工业机器人已经成为现代制造业中最重要的组成部分之一。

工业机器人的应用实例如下：

1）装载机器人。装载机器人将产品放置到目标位置。

2）码垛机器人。码垛机器人可以打包或移动垛台。

3）喷涂机器人。喷涂机器人可对汽车、家具等物体进行喷漆。

4）焊接机器人。焊接机器人可以焊接金属材料。

5）装配机器人。装配机器人主要用于电子产品、汽车等的装配。

6）检测机器人。检测机器人具有视觉识别设备，用于质量检查与控制。

7）制造机器人。制造机器人可进行铸造、磨削、切割、装线及钻孔等加工。

Chapter4 Application of Industrial Robot

Objectives

After reading this chapter, you will be able to:

1) know the main application areas of industrial robot.

2) understand the loading and unloading robot.

3) be familiar with material handing robot.

4) know the main characteristics of fabricating robot.

5) have a basic understanding of the painting, welding and assembling robots.

6) answer the review questions at the end of the chapter.

Reading

Industrial robot can perform various tasks according to the requirements of industrial purpose. Without human supervision, industrial robot can pick and place, paint, weld, and assembly under the programming.

1. Loading and unloading

In modern industry, the loading and unloading must be fast, accurate, stable, and dependable, so as to improve the productivity. The handling efficiency has a significant influence on achieving long-term full capacity operation and improving overall work efficiency. Meanwhile, it should be noted that the robot downtime caused by tool change and scheduled maintenance work should be minimized. Fig. 4-1 indicates the loading and unloading processes which are carried out by an industrial robot.

Fig. 4-1   Loading and unloading

2. Handling

Handling is to move materials and products to the manufacturing plants or continuous processing

between the manufacturing plants. During the handling process, there is no further treatment on the product or the material itself. Thus, manufacturers require fast and low-cost handling labor equipment, for example, a handling robot, as shown in Fig. 4-2.

3. Fabricating

The industrial robot is also applied to treat the metallic materials through casting. Normally, fabricating includes routing, milling, drilling, grinding, polishing, deburring, sanding and riveting. The fabricating robot (Fig. 4-3) has high accuracy and repeatability, which greatly improves the productivity and product quality. Furthermore, the fabricating robot can take human labor out of the dusty and noisy environment.

Fig. 4-2　Materials handling

Fig. 4-3　Fabricating

4. Assembling

Nowadays, assembly robots are mainly used in the fields of circuit boards printing, electric motors, and alternators for automobiles, as shown in Fig. 4-4. All the aforementioned works are cumbersome but with a high assembling requirement. However, the critical problem that limits the automated assembling is the product design. Some products are not designed to assembly by industrial robot, so it requires that products need to be designed to be updated to match with robot assembling.

Fig. 4-4　Assembling

5. Painting

Painting robot (Fig. 4-5), especially of the spray-painting robot, has many advantages. Firstly, it can take the human workers out of the polluted painting conditions caused by the toxic coating materials. Secondly, the painting robot can take more than one end-of arm tool which can handle more types of paints. Lastly, the painting robot can paint with high accuracy, as it can touch the area that may not be directly accessible by humans.

6. Welding

One of the most important uses for robots is to perform spot, stud and other types of welding work. Fig. 4-6 displays a typical welding robot. The welding is a non-stop process once it starts, and the welding area is full of hot gas atmosphere and the smell of molten welding wire. The performance of welding by a robot is much better than human labor. However, the robot is lack of the ability to adjust the welding path or wire feeding as needed, and manual welding can directly make on-site judgments.

Fig. 4-5 Painting

Fig. 4-6 Welding

Vocabulary

purpose /ˈpɜːpəs/ *n.* 目的；用途；意志 *vt.* 决心；企图；打算

supervision /ˌsuːpəˈvɪʒn;ˌsjuː-/ *n.* 监督，管理

load /ləʊd/ *n.* 负载，负荷；装载量 *vt.* 使担负；装填 *vi.* [力] 加载；装载

unload /ʌnˈləʊd/ *vt.* 卸；摆脱…之负担；倾销 *vi.* 卸货；退子弹

accurate /ˈækjərət/ *adj.* 精确的

smooth /smuːð/ *n.* 平滑部分 *adj.* 平稳的 *vt.* 使光滑 *adv.* 光滑地；平稳地

improve /ɪmˈpruːv/ *vt.* 改善，增进；提高…的价值 *vi.* 增加；变得更好

influence /ˈɪnfluəns/ *n.* 影响；势力；感化 *vt.* 影响；改变

workflow /ˈwɜːkˌfləʊ/ *n.* 工作流，工作流程

capacity /kəˈpæsɪtɪ/ *n.* 能力；容量；资格，地位；生产力

prolonged /prəˈlɒŋd/ *adj.* 延长的；拖延的；持续很久的

period /ˈpɪərɪəd/ *n.* 周期，期间；时期；句号 *adj.* 某一时代的

plant /plɑːnt/ *n.* 工厂，车间；植物 *vt.* 种植；培养；安置 *vi.* 种植

rout /raʊt/ *v.* 刻纹；彻底击败，打垮 *n.* 溃败；溃退

mill /mɪl/ *v.* 碾磨，切割（金属），铣 *n.* 磨坊，磨粉厂；机器，铣床

grind /graɪnd/ *n.* 磨 *vt.* 磨碎；磨快 *vi.* 磨碎；折磨

polish /ˈpɒlɪʃ/ *v.* 抛光，擦亮 *n.* 磨光，擦亮；打磨光滑的面

deburr   /diːˈbəː/ v. 抛光，修边；去除毛边

sand   /sænd/ n. 沙；沙地 vt. 擦平或磨光某物

rivet   /ˈrɪvɪt/ n. 铆钉 vt. 铆接；固定；集中于

print   /prɪnt/ n. 印刷业；印刷字体 vt. 印刷；打印 vi. 印刷；出版

circuit   /ˈsɜːkɪt/ n.［电子］电路，回路；巡回 vt. 绕回…环行 vi. 环行

alternator   /ˈɔːltəneɪtə; ˈɒl-/ n.［电］交流发电机

aforementioned   /əˈfɔːmenʃənd/ adj. 上述的；前面提及的

tedious   /ˈtiːdɪəs/ adj. 沉闷的；冗长乏味的

critical   /ˈkrɪtɪk(ə)l/ adj. 鉴定的；［核］临界的；批评的，决定性的；评论的

pollute   /pəˈluːt/ vt. 污染；玷污；败坏

toxic   /ˈtɒksɪk/ adj. 有毒的；中毒的

coating   /ˈkəʊtɪŋ/ n. 涂层；包衣；衣料 v. 给…穿上外衣

spot   /spɒt/ n. 地点；斑点 adj. 现场的 vt. 认出；弄脏

stud   /stʌd/ n. 大头钉；饰钮 vt. 散布；用许多饰钮等装饰

atmosphere   /ˈætməsfɪə/ n. 气氛；大气；空气

molten   /ˈməʊlt(ə)n/ adj. 熔化的；铸造的；炽热的 v. 换毛；脱毛

Notes and analysis

Question 1：Can industrial robot work without human supervision or programming? Why?

Answer：_____

Question 2：What is the role of the material handling robot?

Answer：_____

Question 3：What are the advantages of the fabricating robot?

Answer：_____

Question 4：What is the main reason that limits the application of the automated assembling?

Answer：_____

Question 5：List the reasons of using the painting robot instead of human labor.

Answer：_____

Translation

工业机器人的应用

工业机器人可以根据工业用途的要求进行各种工作。无需人工监督，工业机器人可以根据程序进行装载和放置、喷漆、焊接以及装配等工作。

1. 装载和卸载

在现代工业中，装卸必须快速、准确、平稳、可靠，才能提高生产率。装载效率对实现长时间的全容量操作和提高总工作效率有重要影响。需要注意的是，要尽量减少由工具更换

和预期维护工作引起的机器人停机时间。图 4-1 所示为工业机器人装卸过程。

2. 搬运

搬运是指将物料和产品搬运到制造车间或在制造车间之间进行连续的加工。在搬运过程中，对产品或材料本身不做进一步处理。因此，制造商需要快速低成本的劳动力或设备，如搬运机器人，如图 4-2 所示。

3. 加工

工业机器人也可用于金属材料的切削加工。通常情况下，加工任务包括定径、铣削、钻孔、磨削、抛光、去毛刺、打磨和铆接。加工机器人（图 4-3）具有较高的精度和重复性，提高了生产效率和产品质量。此外，加工机器人可以将人从充满灰尘和噪声的环境中解放出来。

4. 装配

目前，装配机器人主要应用于电路板印制、电机、汽车发电机等领域，如图 4-4 所示。上述工作虽然烦琐，但装配要求较高。但是，限制机器人自动化装配的关键问题是产品设计。有些产品的设计不符合工业机器人装配要求，需要将产品进行设计更新，从而满足装配机器人的要求。

5. 喷涂

喷涂机器人（图 4-5），尤其是喷漆机器人具有许多优点。首先，它可以把人从有毒涂料造成的污染环境中解放出来。其次，喷漆机器人可以携带多个喷头，同时处理多种油漆。最后，喷涂机器人的喷涂精度较高，因为它可以到达到人工可能无法直接到达的区域。

6. 焊接

机器人最大的用途之一是进行点、螺柱和其他类型的焊接工作。图 4-6 所示为典型的焊接机器人。焊接是一个一旦开始就不能间断的过程，焊接区域充满了炽热的气体和熔融的焊丝味道。使用机器人焊接的效果比人工焊接要好得多。然而，机器人缺乏根据需要调整焊接路径或送丝的能力，而人工焊接可以直接做出现场判断。

ROBOT
Chapter 5 Industrial Robot Manufacturers

Objectives

After reading this chapter, you will be able to:

1) know the famous robot manufacturers.
2) describe various characteristics for different robot manufacturers.
3) understand the typical terms used in the manufacturers.
4) be familiar with the major divisions of the manufacturers.
5) answer the review questions at the end of the chapter.

Reading

Almost all the industrial robots are made to complete the manufacturing tasks, such as welding, painting, and assembly work. However, there is no single one robot manufacturer that can dominate the entire industrial business. At present, industrial robots are made by many enterprises mainly in Italy, Switzerland, Germany, Japan, United States, and China, etc.

Although there are a large number of robot manufacturers, the number is gradually decreasing due to company mergers, aequisitions, or bankruptcies. In this chapter, the well-established firms (Fig. 5-1) are discussed here. YASKAWA (Japan), FANUC (Japan), KUKA (Germany), and ABB (Switzerland) are the four world's major suppliers of industrial robots, known as the four major robot families, which take about 50% of the global market share.

a) ABB   b) KUKA   c) YASKAWA   d) FANUC

Fig. 5-1   The products of typical industrial robot manufacturers

Industrial robot usually consists of three parts, including core components, mechanical body and system integration. Core components contain reducer, servo motor and controller, which are the most important parts of the industrial robot industry. The four major manufacturers focus on different technical fields. The core fields of YASKAWA are servo system and motion controller. The core field of FANUC is numerical control system, while KUKA performs well on the control system and mechanical manipulator. ABB's core field is also the control system.

FANUC has three departments. FA (industrial automation) includes Computerized Numerical Control (CNC) system, servo motor, and laser oscillator, etc. Robots include collaborative robot, large and micro robot, arc welding robot and other industrial robots used in various industries. Robomachine (CNC machine tool) includes small machining center, injection molding machine and nanoscale ultra-precision machine tool. The logo of FANUC is shown in Fig. 5-2.

YASKAWA (Fig. 5-3) has four departments, including drive control, robotics, system integration and IT & logistics. Drive control mainly includes servo motor, controller and so on. Roboties includs industrial robot, wafer transfer robot, medical robot, collaborative robot, etc. System integration provides solutions for the application of robot in industries such as steel, papermaking, wastewater treatment, railway transportation, etc.

Fig. 5-2   FANUC Logo   Fig. 5-3   YASKAWA Logo

KUKA (Fig. 5-4) has three departments, including robotics, system integration and Swisslog. The robotics includes six-axis robot, medical robot, palletizing robot and other products. System integration provides solutions for automotive manufacturing, electronic processing, food processing and other industry applications. Swisslog was acquired by KUKA in 2015 to implement advanced automation solutions for hospitals, warehouses and distribution centers. KUKA was acquired by Midea in 2017.

ABB (Fig. 5-5) currently has four main departments: electrical products department, including electric integrated infrastructure, substation, distribution automation, electrical installation, measurement and sensing, and control product; robot and motion control department, including motor, mechanical power transmission device, robot, etc.; industrial automation department, providing integrated solutions and services; power grid department, providing power and automation products, systems, services and software solutions.

Fig. 5-4   KUKA Logo   Fig. 5-5   ABB Logo

Vocabulary

brand /brænd/ v. 加商标于 n. 品牌，商标；类型

enterprise /'entəpraɪz/ n. 企业；事业；进取心；事业心

Italy /'ɪtəli/ n. 意大利

Switzerland /'swɪtsələnd/ n. 瑞士

Germany /'dʒɜːməni/ n. 德国

merge /mɜːdʒ/ vt. 合并；使合并；吞没 vi. 合并；融合

bankruptcy /'bæŋkrʌptsɪ/ n. 破产

establish /ɪ'stæblɪʃ;e-/ v. 建立，创立；确立；获得接受；查实，证实

global /'gləʊb(ə)l/ adj. 全球的；总体的；球形的

consist of 包含；由…组成；充斥着

core /kɔː/ n. 核心；要点；果心；[计] 磁心 vt. 挖…的核

mechanical /mɪ'kænɪk(ə)l/ adj. 机械的；力学的；呆板的；手工操作的

integration /ˌɪntɪ'greɪʃ(ə)n/ n. 集成；综合

reducer /rɪ'djuːsə/ n. [助剂] 还原剂；减径管；减速机

servo motor 伺服马达；伺服电动机

controller /kən'trəʊlə/ n. 控制器；管理员

barrier /'bærɪə/ n. 障碍物，屏障；界线 vt. 把…关入栅栏

division /dɪ'vɪʒ(ə)n/ n. [数] 除法；部门；分配；师（军队）；赛区

CNC system 数控系统

laser /'leɪzə/ n. 激光

oscillator /'ɒsɪleɪtə(r)/ n. [电子] 振荡器

injection /ɪn'dʒekʃ(ə)n/ n. 注射；注射剂；充血；射入轨道

mold /məʊld/ v. 浇铸，塑造 n. 模具；铸模；框架

nanoscale 纳米级

ultra /'ʌltrə/ adj. 极端的，偏激的 n. 过激分子，极端主义者 adv. 很，非常

logistics /lə'dʒɪstɪks/ n. [军] 后勤；后勤学

wafer /'weɪfə/ n. 圆片，晶片；薄片，干胶片；薄饼 vt. 用干胶片封

transfer /træns'fɜː/ v. 转让；转接；移交 n. （地点的）转移

collaborative /kə'læbərətiv/ adj. 合作的，协作的

application /ˌæplɪ'keɪʃ(ə)n/ n. 应用；申请；应用程序

solution /sə'luːʃ(ə)n/ n. 解决方案；溶液；溶解；解答

infrastructure /'ɪnfrəstrʌktʃə/ n. 基础设施；公共建设；下部构造

substation /'sʌbsteɪʃ(ə)n/ n. 分局；变电所；分所；分台

distribution /ˌdɪstrɪ'bjuːʃ(ə)n/ n. 分布；分配；供应

installation /ˌɪnstə'leɪʃ(ə)n/ n. 安装，装置；就职

measurement /ˈmeʒəm(ə)nt/ *n.* 测量；［计量］度量；尺寸；量度制
department /dɪˈpɑːtm(ə)nt/ *n.* 部；部门；系；科；局
transmission /trænzˈmɪʃ(ə)n/ *n.* 传动装置，［机］变速器；传递；传送；播送
grid /ɡrɪd/ *n.* 网格；格子，栅格；输电网
software /ˈsɒf(t)weə/ *n.* 软件

Notes and analysis

Question 1：Is the number of the global robot manufacturers rising? Why?

Answer：_____

Question 2：List the names of the four major industrial robot families.

Answer：_____

Question 3：What are the main parts of the industrial robot?

Answer：_____

Question 4：Is there any difference of the core fields between the manufacturers? What are the main differences?

Answer：_____

Question 5：Conclude the major departments and characteristics of the manufacturers.

Answer：_____

Translation

机器人制造商

几乎所有的工业机器人都是用来完成制造任务的，如焊接、喷漆和装配等。然而，没有一个机器人制造品牌能够包揽整个机器人产业。目前工业机器人的主要生产国有意大利、瑞士、德国、日本、美国和中国等。

虽然机器人制造商数量庞大，但由于公司合并、被收购或破产，机器人制造商的数量在逐步减少。本文仅讨论那些技术成熟的工业机器人制造商（图 5-1）。日本的安川（YASKA-WA）和 FANUC、德国的库卡（KUKA）以及瑞士的 ABB 为全球主要的工业机器人制造商，被称为机器人四大家族，占据全球约 50% 的市场份额。

工业机器人通常由核心零部件、机械本体和系统集成三部分构成。核心零部件包括减速机、伺服电动机和控制器，核心零部件是工业机器人产业的核心壁垒。四大家族在各自技术领域内各有所长，安川的核心领域是伺服系统和运动控制器，FANUC 的核心领域是数控系统，库卡的核心领域是控制系统和机械本体，ABB 的核心领域是控制系统。

FANUC 拥有三大部门。FA（工业自动化）部门包括 CNC 数控系统、伺服电动机和激光振荡器等，机器人部门包括协作机器人、大型与微型机器人、弧焊机器人及其他用于各行各业的工业机器人，数控机床部门包括小型加工中心、注塑机和纳米级超精密机床等。FANUC 的图标如图 5-2 所示。

安川电机（图 5-3）有四大部门，包括驱动控制、机器人、系统集成和 IT 与物流。驱动控制主要包括伺服电动机、控制器等，机器人包括工业机器人、晶圆传送机器人、医疗机器人和协作机器人等，系统集成为机器人应用于钢铁、造纸、废水处理和铁路运输等行业提供解决方案。

库卡（图 5-4）有三个部门，分别为机器人、系统集成和 Swisslog（瑞仕格公司）。机器人包含六轴机器人、医疗机器人和码垛机器人等产品。系统集成提供汽车制造、电子加工、食品加工等行业应用的解决方案。Swisslog 于 2015 年被库卡收购，为医院、仓库和配送中心实施先进的自动化解决方案。2017 年，库卡公司被美的公司收购。

ABB（图 5-5）目前有 4 个主要部门：电气产品事业部，包括电动集成基础设施、变电站、配电自动化、电气安装、测量和传感以及控制等产品；机器人及运动控制部，包括电机、机械动力传动装置、机器人等；工业自动化部，提供集成解决方案与服务；电网事业部，提供电力和自动化产品、系统、服务和软件解决方案。

Chapter 6 Manipulator of Industrial Robot

Objectives

After reading this chapter, you will be able to:

1) understand the components of an industrial robot.
2) know the manipulator.
3) classify the base and arm.
4) be capable of choosing suitable hand.
5) answer the review questions at the end of the chapter.

Reading

Although there are hundreds of robot manufactures world-widely, there are only a few types of structural components for an industrial robot. The three basic parts of an industrial robot are manipulator, controller and power supply.

As one of three basic parts of a robot, a manipulator, as shown in Fig. 6-1, is classified by the arm movement which uses coordination systems to describe. The most common coordination systems include polar coordinate, cylindrical coordinate, Cartesian coordinate, and joint coordinate.

1. Base

The base of a robot is its anchor point, which is designed as a supporting unit for the whole robot. The base is one part of the operating system, and it can perform motion combinations, such as rotation, extension, twisting, and linear motion. The base can be anchored to the floor or the ceiling according to the working re-

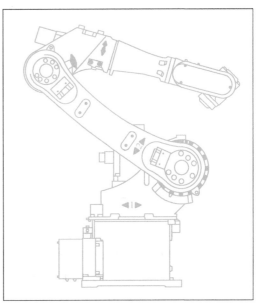

Fig. 6-1　A manipulator

quirement.

2. Arm

Almost all the industrial robots have arms to perform the movements, like grasping and moving the product. The arm is consisted of several axes, such as base, shoulder, elbow, etc. Generally speaking the arm is connected to the base of robot, and it can provide the working envelope per area of floor space. Currently, the demand for six jointed axis industrial robot is very large, and it needs some rather sophisticated computer control.

3. Wrist

The wrist is used to imitate the human wrist, which can perform a wide range of motions. The wrist is connected to a jointed arm with great flexibility, and it can improve the quality of the manufactured product. It can reach the target that is hard for human employee, for instance, in an interior spray painting work or welding inside a pipe.

4. Hand/tool

The specific hand is attached to the end of the wrist, and it is also named end-of-arm tool. The use of hand depends on the production requirement. For example, a gripper (Fig. 6-2) is used to pick, hold, move, and place products, and a sprayer is required in a spray painting line.

Fig. 6-2　A robot hand

As aforementioned, a manipulator is a combination of base, arm, wrist and hand. With the manipulator, the movements including reaching, picking, carrying, moving and placing can be well finished as the design.

Vocabulary

common　　/'kɒmən/ n. 普通；平民；公有地 adj. 共同的；普通的

examine　　/ɪg'zæmɪn;eg-/ vt. 检查；调查；检测；考试 vi. 检查；调查

arm　　/ɑːm/ n. 手臂；武器；袖子；装备；部门 vt. 武装；备战 vi. 武装起来

movement　　/'muːvm(ə)nt/ n. 运动；活动；运转；乐章

coordination　　/kəʊ,ɔː.dɪ'neɪʃən/ n. 协调，调和；对等，同等

polar coordinate　　n. 极坐标

cylindrical coordinate　　n. ［数］柱面坐标

Cartesian coordinate　　n. 笛卡儿坐标

base　　/beɪs/ n. 基底；基础；基地 v. 以……作基础；

anchor　　/'æŋkə/ n. 锚；抛锚停泊 vt. 抛锚；使固定 vi. 抛锚

combination　　/kɒmbɪ'neɪʃ(ə)n/ n. 结合；组合；联合；［化学］化合

rotation　　/rə(ʊ)'teɪʃ(ə)n/ n. 旋转；循环，轮流

extension　　/ɪk'stenʃ(ə)n;ek-/ n. 延长；延期；扩大；伸展；电话分机

twisting　　/'twɪstɪŋ/ n. 快速扭转，缠绕 v. 使弯曲 adj. 曲折的，缠绕的

linear　/'lɪnɪə/ *adj.* 线的，线型的；直线的，线状的；长度的

ceiling　/'siːlɪŋ/ *n.* 天花板；上限

execute　/'eksɪkjuːt/ *vt.* 实行；执行

grasp　/grɑːsp/ *v.* 抓牢，握紧 *n.* 抓，握；理解，把握；权力，控制

shoulder　/'ʃəʊldə/ *n.* 肩，肩膀；肩部 *vi.* 用肩顶 *vt.* 肩负，承担

elbow　/'elbəʊ/ *n.* 肘部；弯头；扶手 *vt.* 推挤；用手肘推开

working envelope　*n.* 工作包络面

sophisticate　/sə'fɪstɪkeɪt/ *adj.* 老于世故的 *v.* 弄复杂；曲解

wrist　/rɪst/ *n.* 手腕；腕关节 *vt.* 用腕力移动

imitate　/'ɪmɪteɪt/ *vt.* 模仿，仿效；仿造，仿制

interior　/ɪn'tɪərɪə(r)/ *n.* 内部，里面；本质 *adj.* 内部的，里面的

pipe　/paɪp/ *n.* 管 *vt.* 用管道输送；用管乐器演奏 *vi.* 吹笛；尖叫

hand　/hænd/ *n.* 手；帮助；指针 *vt.* 传递，交给

component　/kəm'pəʊnənt/ *adj.* 组成的；构成的 *n.* 组成部分；成分；元件

Notes and analysis

Question 1：What is a manipulator?

Answer：_____

Question 2：How many axes of motion does a robot have?

Answer：_____

Question 3：What is a gripper?

Answer：_____

Question 4：How to choose the suitable hand?

Answer：_____

Question 5：What is the role of a base?

Answer：_____

Translation

工业机器人本体

　　尽管世界上有数百家机器人制造商，但是机器人结构部件只有几种。机器人的三个基本部件是工业机器人本体、控制器和电源。

　　工业机器人本体，如图 6-1 所示，是机器人的三个基本部件之一。一般通过臂部动作对本体进行分类，其运动行为则用不同的坐标系统描述。最常用的坐标系统包括极坐标、柱坐标、笛卡儿坐标和关节坐标。

　　1. 底座

　　机器人的底座是机器人的锚点，锚点是整个机器人的支撑单元。底座是操作系统的一部

分，能够进行旋转、伸展、扭转和直线等运动组合。根据工作需要，可将底座固定在地面或天花板上。

2. 手臂

几乎所有的工业机器人都由机械手臂来执行动作，比如抓取和移动产品。手臂由几个轴组成，如底座、肩膀、肘部等。通常而言，手臂与机器人的底座相连，提供工作空间的主截面包络面。目前，六关节轴式工业机器人的需求量很大，它对计算机控制系统的需求也更为复杂。

3. 腕部

机器人本体的腕部是用来模仿人类的手腕的，可用于执行大范围的运动。腕部连接在一个关节臂上，灵活性大，能提高制造产品的质量。工业机器人的腕部可以实现人类难以到达的目标位置，如产品内部的喷漆作业或管道内部的焊接工作。

4. 手/末端执行器

特定的机械予安装在机械人腕部的末端，也称为末端执行器。末端执行器的选择取决于生产的要求。例如夹持器用于挑选、握持、移动和放置产品，而喷漆时则需要使用喷雾器。

如上所述，工业机器人本体由底座、手臂、腕部和末端执行器组成。通过机器人本体的使用，可以很好地完成到达、拾取、搬运、移动和放置等动作。

Chapter**7** Technical Para-meters

Objectives

After reading this chapter, you will be able to:

1) know the technical parameters.
2) understand the degree of freedom.
3) identify the driving mode.
4) classify the working space and working load.
5) be capable of choosing suitable specifications of an industrial robot.
6) answer the review questions at the end of the chapter.

Reading

How to manifest the difference between the numerous industrial robots? The answer is the technical parameters. Diverse robot technical parameters indicate various characteristics, corresponding to their specific application scopes. Industrial robot is a high-precision modern mechanical equipment with numerous technical parameters. Fig. 7-1 presents an example of a teaching industrial robot. Industrial robot mainly includes the following technical parameters.

1. The degree of freedom

The degree of freedom can be explained by the number of axes of the robot. The more axes the robot has, the more degrees of freedom it will have. Furthermore, the robot with more axes owns greater movement flexibility of the mechanical structure. Nevertheless, when the degree of freedom increases, the structure of the robot arm becomes more complex, which causes a negative influence on the robot rigidity. If there are more degrees of freedom on the robot arm than its requirement, the obstacle a-

Fig. 7. 1　A typical industrial robot for education

voidance ability for the robot is provided by the extra degree of freedom. Currently, the degrees of freedom are in the range of 3 to 6, and the selection of degree of freedom depends on the complexity and obstacles of the actual work.

2. Drive mode

Drive mode mainly refers to the power source form of the joint actuator, generally includes hydraulic drive, pneumatic drive, and electric drive. The aforementioned drive modes have individual advantages and features, which has a close relationship with the actual work. The most beneficial point of hydraulic drive is the ability to output a strong driving force with a small driver. For the pneumatic drive, it owns great buffering effect and the ability to get stepless variable speed. The advantages of electric drive are high drive efficiency, easy way to use, and low cost. Nowadays, electric drive mode is commonly used.

3. Control mode

The control mode is also called the control axis mode, and it is mainly used to control the motion trajectory of the robot. Basically, the control mode is classified into two aspects, including servo control and non-servo control. There are two subdivisions of servo control mode, including the continuous trajectory control class and point position control class. Compared with the non-servo controlled robot, the servo controlled robot has a larger memory storage space, which can collect more point addresses to make a multifaceted and stable operation process.

4. Speed

During the operation, working speed is described by the rotation angle or distance of tool center or the mechanical interface center in unit time when the robot moves at a constant speed under a reasonable working load. Normally, a higher maximum speed is the reason of the higher efficiency. Nevertheless, the acceleration and deceleration take more time once the working speed is higher. In another word, it causes high requirements for the maximum rate of acceleration or maximum rate of deceleration of the industrial robot.

5. Workspace

Workspace refers to the maximum range of the origin of end-effector coordinate system in space, or the volume of space occupied by all the points, when the robot operator works normally. The size of the workspace has a close relationship with the size of each link of the robot and the overall structure of the robot. Both the size and shape of the workspace are very important, as the robot may be unable to complete the task due to the blind area which cannot be reached by the hand.

6. Working load

Working load is an important parameter for the industrial robot. During the performance of robot, there is a maximum load which the robot wrist can bear. It is greatly affected by the load mass, working speed and speed acceleration. In the consideration of the safety issues, working load is technically determined as the carrying capacity during high speed operation. Generally, the working load not only refers to the load mass, but also the mass of the robot end-effector.

7. Working accuracy, repetition accuracy and resolution

To make a simple definition, the working accuracy of the robot refers to the error generated by each robot positioning a position. The repetition accuracy refers to the mean value of the error generated by repeatedly positioning a position, while the resolution refers to the minimum moving distance or rotation angle that can be achieved by each axis of the robot. The accuracy of the robot is controlled by these three parameters together.

Vocabulary

technical /ˈteknɪk(ə)l/ *adj.* 工艺的，科技的；技术上的；专门的

parameter /pəˈræmɪtə/ *n.* 参数；系数；参量

manifestation /ˌmænɪfeˈsteɪʃ(ə)n/ *n.* 表现；显示

gap /gæp/ *n.* 间隙；缺口；差距；分歧 *vt.* 使形成缺口 *vi.* 裂开

characteristic /ˌkærəktəˈrɪstɪk/ *n.* 特征；特性；特色 *adj.* 典型的；表示特性的

scope /skəʊp/ *n.* 范围；余地；视野 *vt.* 审视

numerous /ˈnjuːm(ə)rəs/ *adj.* 许多的，很多的

rigidity /rɪˈdʒɪdətɪ/ *n.* ［物］硬度，［力］刚性；严格，刻板；坚硬

extra /ˈekstrə/ *n.* 额外的事物 *adv.* 额外；特别地 *adj.* 额外的；特大的

obstacle /ˈɒbstək(ə)l/ *n.* 障碍，干扰，妨碍；障碍物

avoidance /əˈvɔɪdəns/ *n.* 避免，逃避；废止

drive /draɪv/ *v.* 开车；推动；驱赶；迫使，逼迫

actuator /ˈæktjʊeɪtə/ *n.* ［自］执行机构；激励者；促动器

hydraulic /haɪˈdrɔːlɪk;haɪˈdrɒlɪk/ *adj.* 液压的；水力的；水力学的

pneumatic /njuːˈmætɪk/ *n.* 气胎 *adj.* 气动的；充气的；有气胎的

mode /məʊd/ *n.* 模式；方式；风格；时尚

trajectory /trəˈdʒekt(ə)rɪ;ˈtrædʒɪkt(ə)rɪ/ *n.* ［物］轨道，轨线；［航］［军］弹道

subdivide /ˌsʌbdɪˈvaɪd/ *vt.* 把……再分，把……细分 *vi.* 细分，再分

rotation /rə(ʊ)ˈteɪʃ(ə)n/ *n.* 旋转；循环，轮流

angle /ˈæŋg(ə)l/ *v.* 斜移；谋取 *n.* 角，角度；视角；立场；角铁

distance /ˈdɪstəns/ *n.* 距离；远方；疏远；间隔 *vt.* 疏远

interface /ˈɪntəfeɪs/ *n.* 界面；<计>接口；交界面 *v.* 接合，连接；［计算机］使联系

constant /ˈkɒnst(ə)nt/ *n.* ［数］常数；恒量 *adj.* 不变的；恒定的；经常的

load /ləʊd/ *n.* 负载，负荷；工作量；装载量 *vt.* 使担负；装填

maximum /ˈmæksɪməm/ *n.* ［数］极大，最大限度 *adj.* 最高的；最大极限的

accelerate /ækˈseləreɪt/ *vt.* 使……加快；使……增速 *vi.* 加速；促进；增加

decelerate /diːˈseləreɪt/ *vt.* 使减速 *vi.* 减速，降低速度

index /ˈɪndeks/ *n.* 指标；指数；索引；指针 *vt.* 指出；编入索引中 *vi.* 做索引

repetition /repɪˈtɪʃ(ə)n/ *n.* 重复；背诵；副本

accuracy /ˈækjʊrəsɪ/ *n.* ［数］精确度，准确性

resolution	/rezə'luːʃ(ə)n/ *n.*	[物]分辨率；决议；解决；决心
payload	/'peɪləʊd/ *n.*	有效载荷，有效负荷
capacity	/kə'pæsɪtɪ/ *n.*	能力；容量；资格，地位；生产力
mutually	/'mjuːtʃuəlɪ/ *adv.*	互相地；互助
reinforce	/riːɪn'fɔːs/ *n.*	加强；加固材料 *vt.* 加强，加固 *vi.* 求援；得到增援
vertical	/'vɜːtɪk(ə)l/ *n.*	垂直线，垂直位置 *adj.* 垂直的，直立的
horizontal	/hɒrɪ'zɒnt(ə)l/ *n.*	水平线，水平面 *adj.* 水平的；地平线的
extensibility	/ik,stensə'bɪlətɪ/ *n.*	展开性；可延长性

Notes and analysis

Question 1：List the main driving modes mentioned in this chapter.

Answer：＿＿＿＿＿＿＿＿＿＿＿＿＿＿＿＿＿＿＿＿＿＿＿＿

Question 2：What does the working load mean?

Answer：＿＿＿＿＿＿＿＿＿＿＿＿＿＿＿＿＿＿＿＿＿＿＿＿

Question 3：What is the work envelope?

Answer：＿＿＿＿＿＿＿＿＿＿＿＿＿＿＿＿＿＿＿＿＿＿＿＿

Question 4：Which type of coordinate system produces a teardrop shaped work envelope?

Answer：＿＿＿＿＿＿＿＿＿＿＿＿＿＿＿＿＿＿＿＿＿＿＿＿

Question 5：What is the difference between working accuracy and repetition accuracy?

Answer：＿＿＿＿＿＿＿＿＿＿＿＿＿＿＿＿＿＿＿＿＿＿＿＿

Translation

技 术 参 数

如何区分众多不同的机器人？答案是技术参数。不同的机器人技术参数在特定的应用范围内能够反应出机器人不同的特点。工业机器人是具有很多技术参数的高精密现代机械设备。图 7-1 为一个教学型工业机器人实例。工业机器人主要包括以下七大技术参数。

1. 自由度

自由度可以用机器人的轴数进行解释，机器人的轴数越多，自由度就越多，机械结构运动的灵活性就越大，通用性就越强。但是，自由度增多，会导致机械臂结构变得复杂，降低机器人的刚性。当机械臂的自由度多于完成工作所需要的自由度时，多余的自由度就可以为机器人提供一定的避障能力。目前大部分机器人都具有 3~6 个自由度，可以根据实际工作的复杂程度和障碍进行选择。

2. 驱动方式

驱动方式主要指的是关节执行器的动力源形式，一般有液压驱动、气压驱动和电气驱动。不同的驱动方式有各自的优势和特点，可以根据实际工作的需求进行选择。液压驱动的主要优点在于可以用较小的驱动器输出较大的驱动力。气压驱动的主要优点是具有较好的缓冲作用，可以实现无级变速。电气驱动的优点是驱动效率高，使用方便，而且成本较低。现在比较常用的是电气驱动的方式。

3. 控制方式

机器人的控制方式也被称为控制轴的方式，主要是用来控制机器人运动轨迹。一般来说，控制方式有两种：一种是伺服控制，另一种是非伺服控制。伺服控制方式可以细分为连续轨迹控制和点位控制。与非伺服控制机器人相比，伺服控制机器人具有较大的记忆存储空间，可以存储较多点位地址，使运行过程更加复杂平稳。

4. 工作速度

工作速度指的是机器人在合理的工作载荷之下，匀速运动的过程中，机械接口中心或者工具中心点在单位时间内转动的角度或者移动的距离。简单来说，最大工作速度越快，其工作效率就越高。但是，此时工作速度就要花费更多的时间加速或减速，或者对工业机器人的最大加速度或最大减速度的要求就更高。

5. 工作空间

工作空间指的是机器人正常工作时，末端执行器坐标系的原点能在空间活动的最大范围，或者说该点可以到达所有位置时占的空间体积。工作空间的大小不仅与机器人各连杆的尺寸有关，而且与机器人的总体结构形式有关。工作空间的形状和大小是十分重要的，机器人在执行某作业时可能会因存在末端执行器不能到达的盲区而无法完成任务。

6. 工作载荷

工作载荷是工业机器人的重要参数。工作载荷是指机器人在规定的性能范围内工作时，机器人腕部所能承受的最大负载量。工作载荷不仅取决于负载的质量，而且与机器人运行的速度和加速度的大小和方向有关。为保证安全，将工作载荷这一技术指标确定为高速运行时的承载能力。通常，工作载荷不仅指负载质量，也包括机器人末端执行器的质量。

7. 工作精度、重复精度和分辨率

简单来说，机器人的工作精度是指每次机器人定位一个位置时产生的误差，重复精度是机器人反复定位一个位置产生误差的均值，而分辨率则是指机器人的每个轴能够实现的最小的移动距离或者最小的转动角度。这三个参数共同决定了工业机器人的工作精度。

ROBOT

Chapter8 Robot Coordinates

Objectives

After reading this chapter, you will be able to:

1) classify the robot axis movement.

2) know the definition of coordinate.

3) understand the Cartesian coordinate.

4) identify the cylindrical coordinate.

5) know the polar coordinate.

6) be capable of choosing suitable coordinate.

7) answer the review questions at the end of the chapter.

Reading

To classify the axis movement of the robot, normally there are three coordinate systems, including Cartesian coordinate, cylindrical coordinate, and polar coordinate. The coordinate uses space to describe the movements of the arm.

1. Cartesian Coordinate

Compared with other coordinate systems, Cartesian coordinate is easy to understand. As shown in Fig. 8-1, the main axes are X, Y, and Z. The reference point for the planes is the intersections of all three axes. The centerline of the robot is the reference point for all three axes. The X axis, Y axis, and Z axis are the operating axes. The movement of Z axis is linear, and it can move up and down. The Y axis also performs a linear motion with inside and outside movement. The X axis plays the side to side motion of the manipulator, which can make the manipulator rotate around its base.

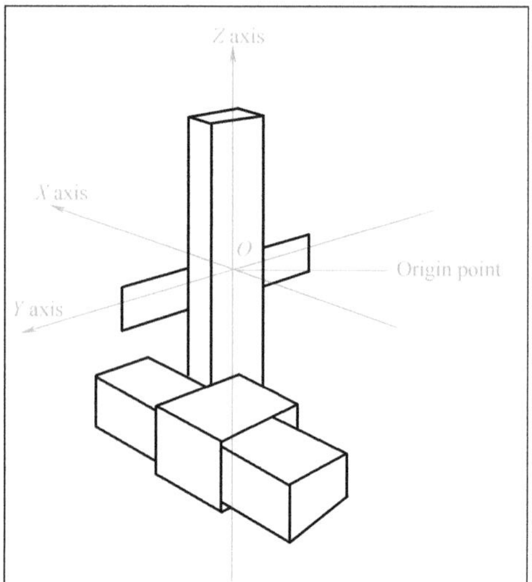

Fig. 8-1　Cartesian coordinate

2. Cylindrical Coordinate

Cylindrical coordinate system (Fig. 8-2) refers to the coordinate system that uses the plane polar coordinate and the Z direction distance to define the spatial coordinate of the object. Same as the spatial rectangular coordinate system, there will be a value variable in the cylindrical coordinate system. The three coordinate variables in the cylindrical coordinate system are r, φ, and z. r is the distance between the projection M' on plane xoy from the origin O to the point M, $r \in [0, +\infty)$. ϕ is an angle rotated from the X-axis to OM' in the counter-clockwise direction from the positive Z-axis, $\varphi \in [0, 2\pi)$. z is the height of the cylinder, $z \in R$.

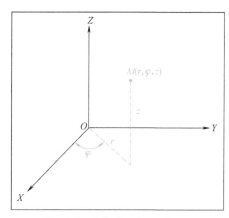

Fig. 8-2   Cylindrical coordinate

3. Polar Coordinate

The polar coordinate system describes a spherical movement mode. Polar coordinate refers to a coordinate system consisting of a pole, a polar axis and a polar diameter in a plane, as shown in Fig. 8-3. A fixed point O on the plane is called the pole. When it puts a ray from O, the polar axis is made. Then it takes a unit of length, and it usually takes the angle to be positive in the counter-clockwise direction. In this way, the position of any point M on the plane can be determined by the length ρ of the line segment OM and the angle θ from OX to OM. The ordered number (ρ, θ) is called the polar coordinate of point M, denoted as $M(\rho, \theta)$; ρ is the polar diameter of M, and θ is the polar angle of M.

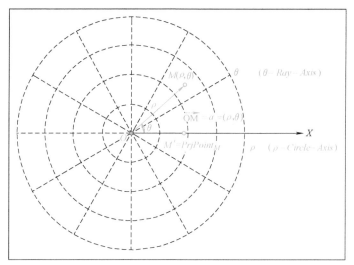

Fig. 8-3   Polar coordinate

4. Coordinate transformation

The coordinates of the polar coordinate system can be converted to the coordinates in the plane rectangular coordinate system. The two coordinates of ρ and θ in the polar coordinate system can be converted to coordinate values in the rectangular coordinate system by the following formula:

$$x = \rho\cos\theta \quad y = \rho\sin\theta \tag{8-1}$$

Coordinate transformation from plane rectangular coordinate system to polar coordinate system can be carried out by calculating the coordinates in polar coordinate system，from the coordinates of x and y in rectangular coordinate system as the following equation：

$$\theta = \arctan(y/x)\,(x \neq 0) \tag{8-2}$$

In the case of $x = 0$, where y is a positive number, $\theta = 90°$ （$\pi/2$ radians）. If y is a negative number, $\theta = 270°$ （$3\pi/2$ radians）.

Vocabulary

Cartesian coordinate　*n.* 笛卡儿坐标

plane　/pleɪn/ *n.* 飞机；平面；水平 *adj.* 平的 *vt.* 刨平 *vi.* 刨

reference　/'ref(ə)r(ə)ns/ *n.* 参考，参照；涉及 证明书 *vt.* 引用 *vi.* 引用

intersection　/ɪntə'sekʃ(ə)n/ *n.* 交叉；十字路口；交集；交叉点

centerline　/'sentəlain/ *n.* 中心线

linear　/'lɪnɪə/ *adj.* 线的，线型的；直线的，线状的；长度的

side　/saɪd/ *n.* 方面；侧面；旁边 *adj.* 旁的，侧的 *vt.* 同意，支持

cylindrical coordinate　*n.* ［数］柱面坐标，圆柱坐标

polar　/'pəʊlə/ *n.* 极面；极线 *adj.* 极地的；两极的；正好相反的

rectangular　/rek'tæŋgjʊlə/ *adj.* 矩形的；成直角的

column　/'kɒləm/ *n.* 纵队，列；专栏；圆柱，柱形物

projection　/prə'dʒekʃ(ə)n/ *n.* 投射；规划；突出；发射；推测

polar coordinates　*n.* ［数］［天］极坐标

spherical　/'sferɪk(ə)l/ *adj.* 球形的，球面的；天体的

pole　/pəʊl/ *n.* 杆；极点；电极 *vt.* 用竿支撑

counterclockwise　/kaʊntə'klɒkwaɪz/ *adj.* 反时针方向的 *adv.* 反时针方向

segment　/'segm(ə)nt/ *n.* 段，部分 *v.* 分割

coordinate transformation　*n.* ［数］坐标变换

convert　/kən'vɜ:t/ *vt.* 使转变；转换…；使…改变信仰 *vi.* 转变，变换

formula　/'fɔ:mjʊlə/ *n.* ［数］公式，准则；配方；婴儿食品

Notes and analysis

Question 1：What is Cartesian coordinate?

Answer：_____

Question 2：List the four classifications of robot coordinate system.

Answer：_____

Question 3：How to realize the coordinate transformation?

Answer：_____

Question 4：How to choose the suitable coordinate system?

Answer：_____

Translation

机器人坐标系

通常用四种坐标系对机器人的轴运动进行分类，包括笛卡儿坐标系、柱坐标系和极坐标系。坐标是用空间来描述手臂运动的。

1. 笛卡儿坐标

与其他坐标系相比，笛卡儿坐标系更易于理解。如图 8-1 所示，主轴是 X 轴、Y 轴和 Z 轴。平面的参考点是三个轴的交点。X 轴、Y 轴和 Z 轴的参考点是机器人的中心线。为了达到目标，X 轴、Y 轴和 Z 轴是运动轴。Z 轴的运动是线性的，可以实现上下运动。Y 轴执行进出方向的线性运动。X 轴带动机械手左右运动，使机械手绕底座旋转。

2. 柱坐标系

柱坐标系（图 8-2）是指使用平面极坐标和 Z 方向距离来定义物体空间坐标的坐标系，与空间直角坐标系相同，柱坐标系中会有一个值变量。柱坐标系中的三个坐标变量是 r、φ 和 z。r 为原点 O 到点 M 在平面 xoy 上的投影 M' 间的距离，$r \in [0, +\infty)$。φ 为从正 Z 轴来看自 X 轴按逆时针方向转到 OM' 所转过的角，$\varphi \in [0, 2\pi)$，z 为圆柱高度，$z \in \mathrm{R}$。

3. 极坐标系

极坐标系统描述了一种球面运动模式。极坐标系是指在平面内由极点、极轴和极径组成的坐标系，如图 8-3 所示。在平面上取定一点 O，称为极点。从点 O 引一条射线 OX，称为极轴。再取定一个单位长度，通常规定角度取逆时针方向为正。这样，平面上任一点 M 的位置就可以用线段 OM 的长度 ρ 以及从 OX 到 OM 的角度 θ 来确定，有序数对 (ρ, θ) 称为点 M 的极坐标，记为 $M(\rho, \theta)$；ρ 称为 M 点的极径，θ 称为 M 点的极角。

4. 坐标转换

极坐标系可以转换为平面直角坐标系。极坐标系中的两个坐标 ρ 和 θ 可以由下面的公式转换为平面直角坐标系下的坐标：

$$x = \rho\cos\theta \quad y = \rho\sin\theta \tag{8-1}$$

平面直角坐标系坐标可转换为极坐标系下的坐标，由直角坐标系的 x、y 坐标计算极坐标系的坐标，如下式所示：

$$\theta = \arctan(y/x)\,(x \neq 0) \tag{8-2}$$

在 $x = 0$ 的情况下：若 y 为正值数，$\theta = 90°$（$\pi/2$ radians）；若 y 为负值，则 $\theta = 270°$（$3\pi/2$ radians）。

ROBOT
Chapter9 Drive System

Objectives

After reading this chapter, you will be able to:

1) understand the role of hydraulics in the operation of robots.

2) identify the word "pneumatics".

3) know the role of electric motors.

4) classify the types of electric motors used in robots.

5) answer the review questions at the end of the chapter.

Reading

The function of an industrial robot is powered by the drive system. To make it clear, the power should be drived in an usable form to move the arm or end-of-arm tools. In other words, the drive system powers the robot and its controller. The main drive systems include hydraulic system, pneumatic system, and electric system.

1. Hydraulic system

The word "Hydraulic" comes from the Greek word, which means water. In the industrial robot drive system, the movements of a manipulator is usually realized by the action of oil. Hydraulic system is much popular due to its wide range of applications. It is capable of picking up, moving the heavy loads, spray painting, and braking of the automobiles.

2. Pneumatic system

The word "Pneu" comes from Latin word "air". Pneumatic is a part of physics that works with air and gases. For the industrial robot, the pneumatics are used to control the end-of-arm tools, such as gripper and absorber, during machining operations. The medium for the pneumatics is the compressed air which is supplied by the air system or compressor air tank. During the operation, the pressure is maintained constantly by the motor. Compared to the hydraulic system, the pneumatic system just uses air which can be directly discharged into the air, while hydraulic system requires a return system to keep the fluids.

3. Electric system

The two main advantages of electric drives are easy maintenance and flexible output functions. The electric motors are controlled by a computer/microprocessor, and it is easy to control the running speed. According to the different current source, electric motors can be classified as direct current (DC) motor and alternating current (AC) motor.

Besides the pneumatic system, hydraulic system, and electric system, stepper motors are also used to demonstrate the basic operations of the industrial robot. Stepper motors are used primarily to change electrical pulses into rotational motion that can be used to produce mechanical movement. The computer then generates the pulses to operate the stepper motor. It is normally adopted in educations, not the modern industrial applications.

Vocabulary

demonstrate /'demənstreɪt/ *vt.* 证明；展示；论证 *vi.* 示威

primarily /'praɪm(ə)rɪlɪ;praɪ'mer-/ *adv.* 首先；主要地，根本上

pulse /pʌls/ *n.* ［电子］脉冲；脉搏 *vt.* 使跳动 *vi.* 跳动，脉跳

rotary /'rəʊt(ə)rɪ/ *n.* 旋转式机器 *adj.* 旋转的，转动的；轮流的

motion /'məʊʃ(ə)n/ *n.* 动作；移动；手势 *vt.* 运动 *vi.* 运动；打手势

education /,edʒʊ'keɪʃn/ *n.* 教育；培养；教育学

hydraulic /haɪ'drɔːlɪk;haɪ'drɒlɪk/ *adj.* 液压的；水力的；水力学的

pneumatic /njuː'mætɪk/ *n.* 气胎 *adj.* 气动的；充气的；有气胎的

electric /ɪ'lektrɪk/ *n.* 电；电气车辆；带电体 *adj.* 电的；电动的；发电的

realize /'rɪəlaɪz/ *vt.* 实现；认识到；了解

brake /breɪk/ *n.* 刹车；阻碍 *v.* 刹车；阻碍

physics /'fɪzɪks/ *n.* 物理学；物理现象

absorber /əb'sɔːbə/ *n.* 减震器；吸收器；吸收体

compressed /kəm'prest/ *adj.* （被）压缩的；扁的 *v.* （被）压紧，精简

compressor /kəm'presə/ *n.* 压缩机；压缩物；［医］压迫器

tank /tæŋk/ *n.* 坦克；水槽；池塘 *vt.* 把…贮放在柜内；打败 *vi.* 乘坦克行进

exhaust /ɪg'zɔːst;eg-/ *n.* 废气；排气管 *v.* 使筋疲力尽；耗尽

fluid /'fluːɪd/ *adj.* 流动的；不固定的，易变的；液压传动的 *n.* 流体，液体

direct current（DC） *n.* ［电］直流电

alternating current（AC） *n.* ［电］交流电

stepper motor *n.* 步进电动机

Notes and analysis

Question 1：List the main types of robot drive systems.

Answer：_____

Question 2：What does the word "hydraulics" mean?

Answer：_____

Question 3：Describe the pneumatic drive motor.

Answer：_____

Question 4：What is a stepper motor used for?

Answer：_____

Question 5：How does DC motor work？

Answer：_____

Translation

驱 动 系 统

工业机器人由驱动系统提供动力。也就是说，要移动机械手臂或手臂末端的工具，需要用不同且合适的动力对其进行驱动。换句话说，驱动系统用来给机器人及其控制器提供动力。驱动系统主要包括液压驱动系统、气动驱动系统和电气驱动系统。

1. 液压驱动系统

"液压"这个词来自希腊语，意思是水。在工业机器人驱动系统中，机械手的运动通常是靠介质油的作用来实现的。液压系统因其应用范围广泛而颇受欢迎。它能够进行装载和移动重型负载，进行喷漆以及汽车制动等工作。

2. 气动驱动系统

"气动"一词来源于拉丁语"air"。气动学是物理学中处理空气和气体的部分。在工业机器人的机械加工过程中，一般通过气动驱动系统来控制夹具和吸收器等手臂末端工具。气动介质是由空气系统或压缩气罐提供的压缩空气。在操作过程中，压力保持恒定。与液压驱动系统相比，气动驱动系统使用空气，可以直接排入空气中，而液压驱动系统则需要一个回流系统来保持液体流动。

3. 电气驱动系统

电气驱动的两个主要优点是易于维护和具有灵活的输出功能。电动机由计算机或微处理器控制，易于控制运行速度。根据电流源的不同，电动机可分为直流电动机和交流电动机。

除了液压驱动系统、气动驱动系统和电气驱动系统外，步进电动机也常用于演示工业机器人的基本操作。步进电动机主要用于将电脉冲转变为旋转运动，从而产生机械运动；然后，计算机产生操作步进电动机所需的脉冲。步进电动机通常用于教学培训，而非现代工业应用。

ROBOT

Chapter 10 Transmission Component

Objectives

After reading this chapter, you will be able to:

1) know the main types of transmission components.

2) understand the roles of gear, belt, and chain.

3) be familiar with the gear.

4) know the differences between various belts.

5) be able to choose proper chain.

6) answer the review questions at the end of the chapter.

Reading

What makes a drive system of an industrial robot? The answer is the transmission component, such as gear, belt, chain, etc. All the mentioned components are used to transfer energy from the actuator, which motivates the arm and related tools.

1. Gear transmission

Gear transmission (Fig. 10-1) is the device which uses gear pairs to transfer motion and power. It is the most widely used mechanical transmission mode in all kinds of modern equipments.

Gear transmission has relatively accurate transmission, high efficiency, compact structure, reliable operation and long service life. It is based on the meshing of tooth-tooth. Gears can be used to increase or decrease the effective speed of the motor. The gear teeth are directly

Fig. 10-1　Gear transmission

involved in the work of the part of gear, so the failure of the gear mainly occurs on the gear teeth. The main failure modes are tooth fracture, tooth surface pitting, tooth surface wear, tooth surface bonding, and plastic deformation.

2. Belt

Belt, also known as transmission belt (Fig. 10-2), is widely used for power transmission in machinery and equipment driven by electric motors and internal combustion engines. According to its usage and structure, belt can be divided into a variety of different categories. Compared with gear transmission and chain transmission, industrial belt transmission shows the advantages of simple mechanism, low noise and low equipment cost, and also it is widely used in various power transmissions.

As a kind of mechanical transmission, belt transmission takes the flexible belt tensioned on the belt wheel to perform the movement or power transmission. Along with different transmission principles, there are frictional belt transmissions driven by the friction between the belt and the wheel, and there are synchronous belt transmissions driven by the meshing of the teeth on belt and wheel. Belt transmission has the characteristics of simple structure, smooth transmission, buffering and absorbing vibration, large shaft spacing and multi-shaft transmission, low cost, no need for lubrication, easy maintenance and so on. There are three main types of belts applied in robots, including V-belts, synchronous belts, and flat belts.

3. Chain

Chain drive is a typical type of transmission mode. It transfers the power and movement of the active sprocket with special tooth shape to the driven sprocket with special tooth shape through the chain, as shown in Fig. 10-3. Chain transmission has many advantages, for the instance, inelastic sliding and slipping phenomenon, accurate average transmission ratio, reliable work, and high efficiency. Under the same working condition, it supplies strong transmission power, good overload capacity, and small transmission size. It can work at high temperature and in humid, dusty, polluted and other harsh environments. The chain transmission also has some shortcomings, for example, high cost, easy to wear, easy to stretch, poor transmission stability, additional dynamic load produced, vibration, impact and noise. It is only used for the transmission between two parallel shafts, and it must not be used in rapid reverse transmission. Roller chains are commonly used in industrial robots.

Fig. 10-2　Belt transmission

Fig. 10-3　Chain transmission

Vocabulary

component　/kəmˈpəʊnənt/ n. 组成部分；成分；组件 adj. 组成的；构成的

gear　/gɪə/ n. 齿轮；传动装置 vt. 开动；搭上齿轮 vi. 适合；搭上齿轮

belt　/belt/ n. 带；腰带；地带 vt. 用带子系住；用皮带抽打

chain　/tʃeɪn/ n. 链；束缚；枷锁 vt. 束缚

energy　/'enədʒɪ/ n. ［物］能量；精力；活力；精神

actuator　/'æktjʊeɪtə/ n. ［自］执行机构；激励者；促动器

transmission　/trænz'mɪʃ(ə)n;-ns-/ n. 传动装置，［机］变速器；传递

compact　/kəm'pækt/ n. 契约 adj. 紧凑的；坚实的 v. 把……压实；使简洁

structure　/'strʌktʃə/ n. 结构；构造；建筑物 vt. 组织；构成；建造

reliable　/rɪ'laɪəb(ə)l/ n. 可靠的人 adj. 可靠的；可信赖的

operation　/ɒpə'reɪʃ(ə)n/ n. 操作；经营；［外科］手术；［数］［计］运算

gear teeth　n. 齿轮齿

mesh　/meʃ/ n. 网眼；网丝；网格 vt. ［机］啮合 vi. 相啮合

failure　/'feɪljə/ n. 失败；故障

fracture　/'fræktʃə/ n. 破裂，断裂 vt. 使破裂 vi. 破裂；折断

pit　/pɪt/ n. 矿井；深坑 vt. 使竞争；使凹下 vi. 凹陷；起凹点

wear　/weə/ v. 穿，戴 n. 衣物；磨损；耐久性

bond　/bɒnd/ n. 结合；约定；黏合剂 vt. 使结合；以…作保 vi. 结合

plastic　/'plæstɪk/ adj. 塑料制的；人造的，塑性的 n. 塑料；塑料学

deformation　/ˌdiːfɔː'meɪʃ(ə)n/ n. 变形

internal　/ɪn'tɜːn(ə)l/ adj. 内部的；本身的；内心的 n. 内脏；内部特征

combustion　/kəm'bʌstʃ(ə)n/ n. 燃烧，氧化；骚动

engine　/'endʒɪn/ n. 引擎，发动机；机车，工具

category　/'kætɪg(ə)rɪ/ n. 种类，分类

tension　/'tenʃ(ə)n/ n. 张力，拉力 vt. 使紧张；使拉紧

wheel　/wiːl/ n. 车轮；转动 vt. 转动；使变换方向 vi. 旋转

frictional　/'frɪkʃənl/ adj. ［力］摩擦的；由摩擦而生的

synchronous　/'sɪŋkrənəs/ adj. 同步的；同时的

buffer　/'bʌfə/ n. ［计］缓冲区；缓冲器，［车辆］减震器 vt. 缓冲

absorb　/əb'zɔːb;-'sɔːb/ vt. 吸收；吸引；承受

vibration　/vaɪ'breɪʃ(ə)n/ n. 振动；犹豫

shaft　/ʃɑːft/ n. 竖井；通风井杆，柄

space　/speɪs/ n. 空间；太空；距离 vt. 隔开 vi. 留间隔

lubrication　/ˌluːbrɪ'keɪʃən/ n. 润滑；润滑作用

V-belts　n. V 带

synchronous belts　n. 同步带

flat belts　n. 平带

sprocket　/'sprɒkɪt/ n. 链轮齿；扣链齿轮

inelastic　/ˌɪnɪ'læstɪk/ adj. 无弹性的；无适应性的；不能适应的

sliding /'slaɪdɪŋ/ n. 滑；移动 adj. 变化的；滑行的 v. 滑动；使滑行

phenomenon /fɪ'nɒmɪnən/ n. 现象；奇迹

overload /əʊvə'ləʊd/ n. 超载量 v. （使）过载，超载

capacity /kə'pæsɪtɪ/ n. 能力；容量；资格，地位；生产力

humidity /hjʊ'mɪdɪtɪ/ n. [气象] 湿度；湿气

dust /dʌst/ n. 灰尘；尘埃 vt. 撒；拂去灰尘 vi. 拂去灰尘；化为粉末

pollution /pə'luːʃ(ə)n/ n. 污染，污染物

harsh /hɑːʃ/ adj. 严厉的；严酷的；刺耳的；粗糙的；刺目的

environment /ɪn'vaɪrənmənt/ n. 环境，外界

shortcoming /'ʃɑːtkʌmɪŋ/ n. 缺点；短处

parallel /'pærəlel/ n. 平行线；对比 adj. 平行的；类似的 vt. 使…与…平行

stretch /stretʃ/ v. 伸展；拉紧 adj. 弹性的，可拉伸的 n. 舒展；伸张

additional /ə'dɪʃ(ə)n(ə)l/ adj. 附加的，额外的

dynamic /daɪ'næmɪk/ n. 动态；动力 adj. 动态的；动力的；动力学的

impact /'ɪmpækt/ vi. 影响；冲突 n. 影响；碰撞 vt. 挤入，压紧

noise /nɔɪz/ n. [环境] 噪音；响声；杂音 vt. 谣传 vi. 发出声音

reverse /rɪ'vɜːs/ v. 颠倒；反转 n. 逆向；相反 adj. 相反的；颠倒的

roller /'rəʊlə/ n. [机] 滚筒；[机] 滚轴；辊子；滚转机

Notes and analysis

Question 1：What makes a drive system of the robot?

Answer：_____

Question 2：What is role of the transmission component?

Answer：_____

Question 3：What are the advantages of gear?

Answer：_____

Question 4：List the types of belts used in the industry.

Answer：_____

Question 5：Describe the differences between the gears and belts.

Answer：_____

Question 6：What are the main shortcomings of a chain transmission?

Answer：_____

Translation

传 动 组 件

工业机器人的驱动系统是由什么组成的？答案是传动部件，如齿轮、带和链等。这些部

件把能量从执行器中传递出来，进而驱动机器人手臂和相关工具。

1. 齿轮传动

齿轮传动（图 10-1）是指由齿轮传递运动和动力的装置，它是现代各种设备中应用最广泛的一种机械传动方式。

齿轮传动比较准确，效率高，结构紧凑，工作可靠，寿命长。齿轮传动是靠齿与齿的啮合进行工作的。齿轮可以用来提高或降低电动机的有效转速。轮齿是齿轮直接参与工作的部分，因此齿轮的失效主要发生在轮齿上。主要的失效形式有轮齿折断、齿面点蚀、齿面磨损、齿面胶合以及塑性变形等。

2. 带传动

传送带就是运用在工业上的皮带，广泛应用于电动机和内燃机驱动的机械和设备上的动力传递，如图 10-2 所示。根据用途与结构不同，带传动可以分为各种不同的类型。与齿轮传动、链条传动相比，工业皮带传动具有机构简单、噪声小和设备成本低等优点，广泛用于各种动力传动。

带传动是利用张紧在带轮上的柔性带进行运动或动力传递的一种机械传动。根据传动原理的不同，有利用带与带轮间的摩擦力传动的摩擦型带传动，也有利用带与带轮上的齿相互啮合传动的同步带传动。带传动具有结构简单、传动平稳、能缓冲吸振、可以在大的轴间距和多轴间传递动力、造价低廉、不需润滑、维护容易等特点。工业机器人中常见的传动带有 V 带，同步带和平带。

3. 链传动

链传动是通过链条将具有特殊齿形的主动链轮的运动和动力传递到具有特殊齿形的从动链轮的一种典型传动方式（图 10-3）。链传动有许多优点，例如，链传动无弹性滑动和打滑现象，平均传动比准确，工作可靠，效率高；相同工况下的传递功率大，过载能力强，传动尺寸小；能在高温、潮湿、多尘、有污染等恶劣环境中工作。链传动的缺点主要有：成本高，易磨损，易伸长，传动平稳性差，运转时会产生附加动载荷、振动、冲击和噪声，仅能用于两平行轴间的传动，不宜用在存在急速反向的传动中。工业机器人中常用的传动链为滚子链。

ROBOT
Chapter 11 Sensors

Objectives

After reading this chapter, you will be able to:

1) know the types of sensors used in the industrial robot.

2) list the types of sensing functions.

3) identify and discuss the key terms used in this chapter.

4) answer the review questions at the end of the chapter.

Reading

To act as human, the five senses including sight, hearing, taste, touch and smell must be considered. Fig. 11-1 shows the touching sense of a human-like robot. For an industrial robot, the sensors are used to realize the sensing functions, in which the vision is the most important part.

In order to be able to locate and recognize products, the presence, size and shape of products should be "felt" by the robot, so a transducer which converts nonelectrical energy into electrical energy serves as a sensor.

In robot manufacture, sensor is an essential part, which is used to measure the positions

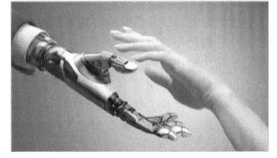

Fig. 11-1　Touching sense of a human-like robot

and speed of tool and products and even indicate the limits of stroke on individual joints within the robot system. For example, a sensor is incorporated at the end of a conveyor to sense the presence of the product at the stop position, as shown in Fig. 11-2.

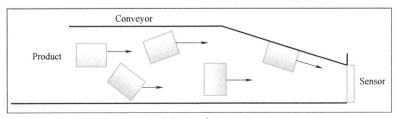

Fig. 11-2　Product position

Normally, the sensor is classified as contact or noncontact. Existence, size, pressure, temperature, and tactile sensors all respond to contact catalogs. Pressure change, temperature change, and electromagnetic change can all be sensed via noncontact methods.

1. Temperature sensing

Temperature sensing is very common. The temperature change is sensed by the temperature sensor. When the temperature increases, the resistance of the thermistor decreases, so the current in the circuit decreases, and vice versa.

2. Displacement sensing

The displacement sensing is important to describe the location information to the computer or microprocessor. As shown in Fig. 11-3, in order to provide a displacement message, a resistive sensor which has a fixed resistance with a slider contact can be used. The change in the resistance with the applied force causes the current variation in the circuit, which helps the computer to act and finally determine the exact displacement of the arm or the end-of-arm tools.

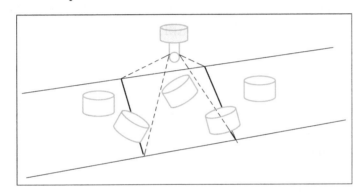

Fig. 11-3　Displacement sensing

3. Speed sensing

Speed sensing is adopted to detect the speed of the motor, in which the tachometer and the photocell are implemented. Speed sensing is also used in the motion control of the industrial robot.

Vocabulary

sight　/saɪt/ n. 视力；景象 adj. 即席的 vt. 看见 vi. 瞄准；观看

hearing　/'hɪərɪŋ/ n. 听力；审讯，听讯 v. 听见

taste　/teɪst/ n. 味道；品味；审美 vt. 尝；体验 vi. 尝起来

smell　/smel/ n. 气味，嗅觉 v. 嗅，闻；察觉到

invariably　/ɪn'veərɪəblɪ/ adv. 总是；不变地；一定地

essential　/ɪ'senʃ(ə)l/ n. 本质；要素；必需品 adj. 基本的；必要的

velocity　/və'lɔsəti/ n. ［力］速率；迅速；周转率

stroke　/strəuk/ n. 冲程；笔画；打击 vt. 抚摸；敲击 vi. 击球

individual　/ˌɪndə'vɪdʒuəl/ adj. 单独的，独特的 n. 个人，个体

joint　/dʒɔɪnt/ adj. 联合的，连接的 n. 关节；接合点 v. 连接，贴合

presence /'prez(ə)ns/ *n.* 存在；出席；参加；风度；仪态

size /saɪz/ *n.* 大小；尺寸 *adj.* 一定尺寸的 *vt.* 依大小排列 *vi.* 可比拟

shape /ʃeɪp/ *n.* 形状；模型 *vt.* 形成；塑造 *vi.* 形成；成形；成长

transducer /trænz'djuːsə;-ns-/ *n.* ［自］传感器，［电子］变换器

contact /'kɒntækt/ *n.* 接触，联系 *vt.* 使接触，联系 *vi.* 使接触，联系

noncontact /ˌnɒn'kɒntækt/ *n.* 无触头，无触点 *adj.* 没有接触的

touch /tʌtʃ/ *v.* 接触；轻按；轻弹；相互接触 *n.* 触碰；轻按；触觉

force /fɔːs/ *n.* 力量；武力；军队 *vt.* 促使，推动；强迫；强加

pressure /'preʃə/ *n.* 压力；压迫，［物］压强 *vt.* 迫使；密封；使……增压

temperature /'temprətʃə(r)/ *n.* 温度

tactile /'tæktaɪl/ *adj.* ［生理］触觉的，有触觉的；能触知的

magnetic /mæg'netɪk/ *adj.* 地磁的；有磁性的；有吸引力的

pattern /'pæt(ə)n/ *n.* 模式；图案；样品 *vt.* 模仿；以图案装饰 *vi.* 形成图案

thermistor /θɜː'mɪstə/ *n.* ［电子］热敏电阻；电热调节器

vice versa /ˌvaisi'vɜːsə/ *adv.* 反之亦然

resistive /rɪ'zɪstɪv/ *adj.* 有抵抗力的；抗……的，耐……的；电阻的

slider /'slaɪdə/ *n.* 滑动器，滑竿；浮动块，滑动块

resistance /rɪ'zɪst(ə)ns/ *n.* 阻力；电阻；抵抗；反抗；抵抗力

tachometer /tæ'kɒmɪtə/ *n.* 转速计，转速表

photocell /'fəʊtəʊsel/ *n.* ［电］光电池；［电子］光电管

Notes and analysis

Question 1：What are the five senses for human?

Answer：_____

Question 2：List the two types of the sensor.

Answer：_____

Question 3：List the devices used to sense touch, force, pressure, temperature, and vision for an industrial robot.

Answer：_____

Question 4：What are the two types of devices that sense temperature?

Answer：_____

Question 5：What is a machine vision system and why is it needed?

Answer：_____

Translation

传　感　器

人类具有视觉、听觉、味觉、触觉和嗅觉五种感官。图 11-1 所示为类人机器人的触觉

传感器。对于工业机器人而言，传感器是用来实现感知功能的，其中视觉是最重要的。

为了能够定位并识别产品，机器人需要"感知"产品的存在、大小和形状。能够将非电信号转换成电信号的元件便是传感器。

传感器是机器人制造生产应用中的重要组成部分。在机器人系统中，传感器通常用于测量机械手臂的位置和速度，或指示单个关节的行程限制。如图11-2所示，传感器安装在传送带的末端，用于感知产品在停止位置的存在。

传感器一般分为接触式和非接触式。存在、大小、压力、温度和触觉传感器都对接触做出反应。压力变化、温度变化和电磁变化都可以用非接触方法探测到。

1. 温度传感器

温度传感器是一种非常常见的传感器，它可以检测到温度的变化。当温度升高时，热敏电阻减小，因此电路中的电流减小，反之亦然。

2. 位移传感器

位移传感器是向计算机或微处理器传递位置信息的重要手段。如图11-3所示，为了给出位移信息，可以使用具有固定电阻和滑块接触的电阻传感器。电阻随外加力的变化引起电路中电流的变化，从而在计算机的计算分析下，最终确定机械手臂或手臂末端工具的准确位移。

3. 速度传感器

采用速度传感器和光电管可对电动机的转速进行检测。速度传感器也可以用于工业机器人的动作控制。

Industrial Robot

ROBOT
Chapter 12 Robot-Computer Interface

Objectives

After reading this chapter, you will be able to:

1) understand the relationship between the robot and computer interface.

2) know the software used in robot.

3) identify and discuss the key terms.

4) answer the review questions at the end of the chapter.

Reading

Robot does not work alone as it always interacts with conveyors, transfer lines, and other material-handling equipment (Fig. 12-1). In order to make the robot work safely and accurately, the interconnection to other equipment is necessary, normally under the constructions of controller or microprocessor. This chapter introduces some relative basic terms.

1. Robot-computer interface

To accomplish the connection of a computer and a robot, it integrates a number of configurations or systems. There are three major parts of the microcomputer, including memory, central processing unit (CPU), and input/output (I/O) ports.

Fig. 12-1   Robot-computer interface

2. Memory

The function of the memory is to store the data. The data includes the part which is processed by CPU and the feedback data from the processing.

3. Central processing unit

Central processing unit consists of the control circuitry, arithmetic logic unit (ALU), register, and address program counter, which is also called micro-processing (MPU). The code from the memory enters into the CPU and then is decoded and executed. The central processing unit can work quickly and respond to instruction without slowing down the process.

4. Input/output（I/O）ports

The function of I/O ports is to connect the interface with the outside part. The input ports like the keyboards, cameras and other devices supply the data to the CPU. The output port is responsible to send data to an output device to execute the movements.

5. Language

The languages are quite different between the robot manufacturers, which makes the language unique for a particular controller. Basically, the robots are preprogrammed according to the requirement of users, while some of them use languages, such as VAL, HELP, AML, MOL, RPL, and RAIL.

6. Software

The program that controls the robot is not synchronized with computers. The use of the robot is not influenced when the programming is going on. In a robot system, the software is adopted to check the status of the program. Before the equipment is debugged or sold, the software must be debugged, and the working program must be edited. In addition, the use of software can save a lot of time. When the new program is made, the robot can remain in work.

Vocabulary

interact /ˌɪntərˈækt/ *n.* 幕间剧；幕间休息 *vt.* 互相影响；互相作用

conveyor /kənˈveɪə/ *n.* 输送机，[机] 传送机；传送带；运送者，传播者

interconnection /ˌɪntəkəˈnekʃən/ *n.* [计] 互连；互相联络

construction /kənˈstrʌkʃ(ə)n/ *n.* 建设；建筑物；解释；造句

controller /kənˈtrəʊlə/ *n.* 控制器；管理员；主计长

microprocessor /ˌmaɪkrə(ʊ)ˈprəʊsesə/ *n.* [计] 微处理器

aggregate /ˈægrɪgeɪt/ *n.* 合计；集合体 *adj.* 聚合的 *v.* 集合；聚集；合计

accomplish /əˈkʌmplɪʃ;əˈkɒm-/ *vt.* 完成；实现；达到

mate /meɪt/ *n.* 助手 *vt.* 使配对；使一致；结伴 *vi.* 成配偶；紧密配合

memory /ˈmem(ə)rɪ/ *n.* 记忆，记忆力；内存，[计] 存储器；回忆

central processing unit *n.* [计] 中央处理机；中央处理单元

input/output（I/O） *n.* 输入/输出

port /pɔːt/ *n.* （计算机的）端口；左舷；舱门 *vt.* 持（枪）

store /stɔː/ *n.* 商店；储备，贮藏；仓库 *vt.* 贮藏，储存

data /ˈdeɪtə/ *n.* 数据；资料

circuitry /ˈsɜːkɪtrɪ/ *n.* 电路；电路系统；电路学；一环路

arithmetic logic unit（ALU） *n.* 算术逻辑单元

register /ˈredʒɪstə/ *n.* 登记表；声区；套准 *v.* 登记

counter /ˈkaʊntə/ *n.* 柜台；计数器 *vi.* 逆向移动，对着干；反驳 *adj.* 相反的

code /kəʊd/ n. 代码，密码；编码；法典 vt. 编码；制成法典 vi. 指定遗传密码

recall /rɪˈkɔːl/ n. 召回；回忆；撤销 vt. 召回；回想起，记起；取消

slowdown /ˈsləʊdaʊn/ n. 减速；怠工；降低速度

keyboard /ˈkiːbɔːd/ n. 键盘 vt. 键入 vi. 用键盘进行操作

device /dɪˈvaɪs/ n. 装置；策略；图案；设备；终端

language /ˈlæŋgwɪdʒ/ n. 语言；语言文字；表达能力

unique /juːˈniːk/ n. 独一无二的 adj. 独特的，稀罕的；[数] 唯一的

software /ˈsɒf(t)weə/ n. 软件

status /ˈsteɪtəs/ n. 地位；状态；情形

debug /diːˈbʌg/ vt. 调试；除错，改正有毛病部分

remain /rɪˈmeɪn/ n. 遗迹；剩余物，残骸 vi. 保持；依然；留下；残存

synchrouized /ˈsɪŋkrənaɪzd/ adj. 同步的；同步化的

Notes and analysis

Question 1：What is the purpose of a controller?

Answer：_____

Question 2：What is a MPU？

Answer：_____

Question 3：List six languages used by the industrial robot.

Answer：_____

Question 4：What can the software do？

Answer：_____

Question 5：Identify the robot-computer interface.

Answer：_____

Translation

机器人计算机界面

机器人不是单独工作的，因为它总是与传送带、输送线和其他产品处理设备相互作用。为了使机器人安全、准确地工作，通常用控制器或微处理器与其他设备进行互联。本章将介绍一些相关的基本术语。

1. 机器人计算机界面

机器人计算机界面集合了多种配置或系统，用于完成计算机和机器人的联系。微型计算机包括三个主要部分，分别为内存、中央处理器（CPU）和输入/输出（I/O）端口。

2. 内存

内存的功能是存储数据。数据包括 CPU 处理的数据和处理后的反馈数据。

Industrial Robot

3. 中央处理器

中央处理器由控制电路、算术逻辑单元（ALU）、寄存器和地址程序计数器（也称为微处理器）组成。内存中的代码进入 CPU 后，进行解码并执行。中央处理器能够快速工作，响应指令，使过程不减速。

4. 输入/输出（I/O）端口

I/O 端口的功能是将接口与外部连接起来。输入端口（如键盘、相机和其他设备）向 CPU 提供数据。输出端口负责根据输出设备发送数据来执行动作。

5. 语言

机器人制造商之间的语言是非常不同的，这使得特定控制器的语言是独一无二的。基本上，这些机器人是根据用户的需求进行预编程的，而其中一些机器人使用 VAL、HELP、AML、MOL、RPL 和 RAIL 等语言。

6. 软件

控制机器人的程序与计算机是不同步的。在使用软件进行编程的过程中，不会影响到机器人的正常使用。在机器人系统中，软件用来检查程序的状态。在设备调试或设备出售前，必须调试好软件、编辑好工作程序。此外，使用软件可以节省很多时间。当编制新的程序时，机器人可以不受影响继续工作。

ROBOT
Chapter**13** Program and Programming

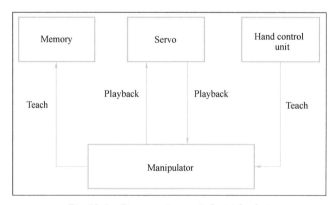

Objectives

After reading this chapter, you will be able to:
1) understand the roles of program and programming.
2) know the different programming methods.
3) recognize the differences between the program languages.
4) classify the program structure.
5) answer the review questions at the end of the chapter.

Reading

A robot will not work on its own unless it receives a task. The robot program is used to make the robot complete a specific task, while robot programming is to make the robot complete the task by setting a certain sequence of actions, as shown in Fig. 13-1. The task operation mainly includes motion instructions and operation instructions, both of which are controlled by the programs.

Fig. 13-1 Programming an industrial robot

At present, the commonly used robot programming methods include teaching programming and offline programming. Teaching programming is practical and easy to operate, which includes teaching, editing and track reproduction. Based on computer graphics, the off line programming establishes the geometric model and obtains the job planning trajectory through a programming algorithm.

The main difference between teaching programming and offline programming is that offline programming does not affect the field work of the robot.

According to the different requirements of the robot, different programming is required. There is a great relationship between the programming ability and the programming method, which determines the adaptability and operating ability of the robot. With the widely spread application of computers in industry, the computer programming of an industrial robot has become more and more important. Robot programming inevitably involves the robot language. Robot language is a method of using symbols to describe robot actions. It makes the robot perform various operations according to the intention of the programmer through the descriptions. Robot programming language is used to describe the operations that can be performed by robots. An available robot programming language should include the instruction set, the program format and structure, the program expression code, and carrier. Depending on the language level, the number of instructions can range from several to dozens, and also the simpler the better. The format and structure of the program are the key parts, which should be universal. The program expression code and carrier are used to transfer the source programs.

There are more than 1000 programming languages in the world, and there are several popular programming languages in robotics today, each of which has different advantages for robots.

1) C and C ++ are common languages for many hardware libraries. C language can be compiled in a simple way, and it can process the low-level memory. C++ language, with its interactivity and cross-platform features, is a very mature programming language. Fig. 13-2 introduces two typical C and C++ textbooks.

Fig. 13-2   C/C++ textbooks

2) Python (Fig. 13-3) receives a great revival in recent years especially in robotics. The main focus of the Python language is ease of use. Python allows simple binding using C/C ++ code.

3) AL is a high-level programming language that describes tasks such as assembly. It is a compiler that converts program into machine code. AL language has a great influence on other languages and plays a leading role in general robot languages.

4) AML is an interactive task-oriented programming language specifically used to control the manufacturing processes (including robots). It supports position and attitude demonstration, joint interpolation motion, linear motion, continuous trajectory control, and force perception, which pro-

Fig. 13-3　Python textbooks

vides robot motion and sensor instruction, communication interface and strong data processing function (capable of group operation of data).

5) MCL is a robot language that is developed for off line programming of an unit of work. The unit of work can be a variety of robots and peripherals, CNC machines, tactile and visual sensors.

Vocabulary

initiative　/ɪˈnɪʃɪətɪv;-ʃə-/ *n.* 主动权；新方案 *adj.* 主动的；自发的；起始的

programming　/ˈprəugræmɪŋ/ *n.* 设计，规划；编制程序，[计] 程序编制

sequence　/ˈsiːkw(ə)ns/ *n.* [数] [计] 序列；顺序；续发事件 *vt.* 按顺序排好

instruction　/ɪnˈstrʌkʃ(ə)n/ *n.* 指令，命令；指示；教导；用法说明

offline　/ɒfˈlaɪn/ *adj.* （计算机）未联网的；离线的 *adv.* 未连线地；脱机地

graphic　/ˈɡræfɪk/ *adj.* 形象的；图表的；绘画似的

geometric　/ˌdʒɪəˈmetrɪk/ *adj.* 几何学的；[数] 几何学图形的

trajectory　/ˈtrædʒɪkt(ə)rɪ/ *n.* [物] 轨道，轨线；[航] [军] 弹道

algorithm　/ˈælɡəˈrɪðəm/ *n.* [计] [数] 算法，运算法则

field work　*n.* 现场工作

mode　/məud/ *n.* 模式；方式；风格；时尚

adaptability　/əˌdæptəˈbɪlətɪ/ *n.* 适应性；可变性；适合性

language　/ˈlæŋɡwɪdʒ/ *n.* 语言；语言文字；表达能力

programmer　/ˈprəugræmə/ *n.* [自] [计] 程序设计员

expression　/ɪkˈspreʃən/ *n.* 表现，表示，表达；表情，态度

code　/kəud/ *n.* 代码，密码；编号；法典 *vt.* 编码 *vi.* 指定遗传密码

dozen　/ˈdʌz(ə)n/ *n.* 十二个，一打 *adj.* 一打的

format　/ˈfɔːmæt/ *n.* 格式；版式；开本 *vt.* 使格式化；规定…的格式 *vi.* 设计版式

universal　/juːnɪˈvɜːs(ə)l/ *n.* 一般概念；普遍性 *adj.* 普遍的；通用的；全体的

hardware　/ˈhɑːdweə/ *n.* 计算机硬件；五金器具

interactivity　/ˈɪntəˈæktɪv/ *n.* 交互性；互动性

feature　/ˈfiːtʃə(r)/ *n.* 特色，特征 *vt.* 特写；以…为特色 *vi.* 起重要作用

revival　/rɪˈvaɪvl/ *n.* 复兴；复活；苏醒；再生效

attitude /'ætɪtjuːd/ n. 态度；看法；意见；姿势

demonstration /ˌdemən'streɪʃ(ə)n/ n. 示范；证明；示威游行

joint /dʒɔɪnt/ adj. 联合的；连接的 n. 关节；接缝 v. 连接，接合

interpolation /ɪnˌtɜːpə'leʃən/ n. 插入；篡改；填写；插值

perception /pə'sepʃ(ə)n/ n. 认识能力；知觉，感觉；洞察力

peripheral /pə'rɪfərəl/ adj. 外围的；次要的；（神经）末梢区域的 n. 外部设备

CNC n. 电脑数值控制（Computer Numerical Control）

tactile /'tæktaɪl/ adj. ［生理］触觉的，有触觉的；能触知的

otes and analysis

Question 1：How does the robot work?

Answer：_____

Question 2：What is the program?

Answer：_____

Question 3：Describe the teaching programming.

Answer：_____

Question 4：What is the main difference between the teaching programming and offline programming?

Answer：_____

Question 5：What does the programming language include?

Answer：_____

Question 6：How many languages are mentioned in this chapter? What are they?

Answer：_____

Translation

机器人编程

机器人不会自己主动工作，除非它收到了工作任务。机器人程序用于使机器人完成特定的工作任务，机器人编程则是通过设置一定的动作顺序使机器人完成该任务，如图 13-1 所示。任务运行主要包括运动指令和作业指令，两者均由程序控制。

目前，常用的机器人程序编制包括示教编程和离线编程两种。示教编程实用性强、操作简便，通过示教、编辑和轨迹再现实现。离线编程方法基于计算机图形学，建立几何模型，通过规划算法来获取作业规划轨迹。示教编程和离线编程的主要区别是，离线编程不影响机器人的现场工作。

对机器人而言，不同的工作要求需要不同的编程。编程能力和编程方式有很大的关系，编程方式决定着机器人的适应性和作业能力。随着计算机在工业上的广泛应用，工业机器人的计算机编程变得日益重要。机器人编程必然涉及机器人语言。机器人语言是使用符号来描

述机器人动作的方法，它通过对机器人的描述，使机器人按照编程者的意图进行各种操作。机器人编程语言用来描述可被机器人执行的作业操作，一个可用的机器人编程语言应包括指令集合、程序格式与结构以及程序表达码和载体。随语言水平不同，指令个数可由数个到数十个，越简单越好。程序的格式与结构是关键部分，应有通用性。程序表达码和载体则用予传递源程序。

世界上有超过 1000 种编程语言，目前机器人技术中流行的编程语言有以下几种，每种语言对机器人有不同的优势。

1）C 和 C++是很多硬件库的通用语言。C 语言能以简易的方式编译、处理低级存储器。C++语言具有交互性，具有跨平台的特性，是一门非常成熟的编程语言。图 13-2 所示为常用的 C 和 C++指导书。

2）Python（图 13-3）近年来尤其在机器人技术方面出现了巨大的复苏。Python 语言的主要特点是十分易用。Python 允许使用 C/C++代码进行简单的绑定。

3）AL 语言是一种高级程序设计系统，可描述诸如装配一类的任务。它能够将程序转换为机器码的编译程序。AL 语言对其他语言有很大的影响，在一般机器人语言中起主导作用。

4）AML 语言是一种交互式、面向任务的编程语言，专门用于控制制造过程（包括机器人）。它支持位置和姿态示教、关节插补运动、直线运动、连续轨迹控制和力觉，可向机器人提供运动和传感器指令、通信接口和很强的数据处理功能（能进行数据的成组操作）。

5）MCL 语言是为工作单元离线编程而开发的一种机器人语言。工作单元可以是各种形式的机器人及外围设备、数控机械、触觉和视觉传感器。

Chapter **14** Machine Vision

Objectives

After reading this chapter, you will be able to:

1) have an understanding of the machine vision.

2) know the difference between machine vision and human eyes.

3) understand the role of machine vision in industry.

4) be familiar with the knowledge of industrial camera.

5) classify the main property of industrial camera.

6) answer the review questions at the end of the chapter.

Reading

Instead of human eyes, machine uses machine vision to carry out the measurement and judgment. The machine vision system refers to the conversion of the ingested targets into image signals through the machine vision products. Then, the image signals are processed to an image processing system and converted into digital signals according to the pixel distribution, brightness, color and other information. The signals are treated via certain operations by the image system to extract the features. Thus, further instruction of the field equipment action can be made according to the results of the discrimination.

As shown in Fig. 14-1, a typical industrial machine vision system consists of light source, lens, camera, image processing, image processing software, monitor, input and output units, etc.

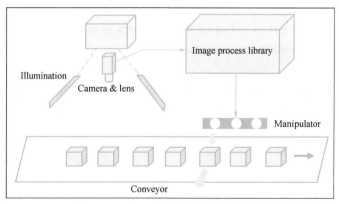

Fig. 14-1 Machine vision system

Compared with human vision, machine vision has the following advantages. Firstly, non-contact detection will not imitate the observer and the observed object. Secondly, it can work stably for a long time. Thirdly, it owns high measuring accuracy. Lastly, the camera speed automatically matches the speed of the measured object to obtain an ideal image.

As industrial camera is a key component of the machine vision system. The basic function of the camera is to transform optical signal into ordered electrical signal. Therefore, selecting an appropriate industrial camera is also an important link in the design of a machine vision system. Industrial camera not only directly determines the resolution and quality of images collected, but also directly relates to the operation mode of the entire system.

The main parameters of industrial cameras are listed as follows.

1. Resolution

Resolution is the number of pixels of each image collected by the camera. For industrial digital cameras, it corresponds directly to the pixel number of the photoelectric sensor. For industrial digital analog camera, it depends on the video format.

2. Pixel depth

Pixel depth is the number of bits of data per pixel. It is generally 8Bit, and there are also 10Bit and 12Bit for industrial digital cameras.

3. Max frame rate/line frame

The rate of image collection and transmission by the camera is generally the number of frames per second for the face-array camera and the number of lines per second for the line-array camera.

4. Exposure mode and shutter speed

For industrial line array cameras, the mode of line-by-line exposure is adopted. Fixed line frequency and external trigger synchronization acquisition mode can be selected. The exposure time can be consistent with the line cycle, or a fixed time can be set. The array camera has several common modes such as frame exposure, field exposure and rolling line exposure. Industrial digital cameras generally provide the ability to trigger color images externally. The shutter speed can be up to 10μs, and high-speed cameras can be faster.

5. Pixel size

Pixel size and pixel number (resolution) jointly determine the size of the camera's target surface. At present, the pixel size of industrial digital camera is generally 3~10μm. The smaller the pixel size is, the more difficult it is to manufacture.

6. Spectral response characteristics

It refers to the sensitivity characteristics of the pixel sensor to different light waves. The general response range is from 350~1000nm. Some cameras will add filters in front of the lens to filter out infrared light.

Vocabulary

vision /ˈvɪʒ(ə)n/ n. 视力；美景；眼力；想象力 vt. 想象；显现；梦见

14

judgment /ɪdʒʌdʒmənt/ n. 判断；裁判；判决书；辨别力

community /kəˈmjuːnətɪ/ n. 社区；[生态] 群落；共同体；团体

dedicate /ˈdedɪkeɪt/ vt. 致力；献身；题献

pixel /ˈpɪks(ə)l;-sel/ n. （显示器或电视机图像的）像素

distribution /dɪstrɪˈbjuːʃ(ə)n/ n. 分布；分配；供应

brightness /ˈbraɪtnɪs/ n. [光] [天] 亮度；聪明，活泼；鲜艳；愉快

color /ˈkʌlə(r)/ n. 颜色；肤色；颜料 vt. 粉饰；歪曲 vi. 变色；获得颜色

information /ɪnfəˈmeɪʃ(ə)n/ n. 信息，资料；知识；情报

extract /ˈekstrækt;ɪkˈstrækt/ n. 摘录；汁 v. 提炼；选取，摘录；取出

discrimination /dɪˌskrɪmɪˈneɪʃ(ə)n/ n. 歧视；区别，辨别；识别力

lens /lenz/ n. 透镜，镜头；眼睛中的水晶体；晶状体 vt. 给……摄影

camera /ˈkæm(ə)rə/ n. 照相机；摄影机

monitor /ˈmɒnɪtə/ n. 监视器；监听器；监控器；显示屏；班长 vt. 监控

imitation /ɪmɪˈteɪʃ(ə)n/ n. 模仿，仿造；仿制品 adj. 人造的，仿制的

observer /əbˈzɜːvə/ n. 观察者；[天] 观测者；遵守者

spectral /ˈspektr(ə)l/ adj. [光] 光谱的；幽灵的；鬼怪的

collation /kəˈleɪʃ(ə)n/ n. 校对

ideal /aɪˈdɪəl;aɪˈdiːəl/ n. 理想；典范 adj. 理想的；完美的；想象的

appropriate /əˈprəʊprɪeɪt/ adj. 适当的；恰当的；合适的 vt. 占用，拨出

link /lɪŋk/ n. [计] 链环，环节 vt. 连接；联合 vi. 连接起来

resolution /rezəˈluːʃ(ə)n/ n. [物] 分辨率；决议；解决；决心

photoelectric /ˌfəʊtəʊɪˈlektrɪk/ adj. [电子] 光电的

video /ˈvɪdɪəʊ/ n. [电子] 视频；录像，录像机 adj. 视频的 v. 录制

format /ˈfɔːmæt/ n. 格式；版式 vt. 使格式化；规定…的格式 vi. 设计版式

pixel depth n. 像素深度

frame /freɪm/ n. 框架；结构；adj. 有构架的 vt. 给（图画或照片）配框；设计

array /əˈreɪ/ n. 数组，阵列；列阵；一系列；衣服 vt. 排列，部署；打扮

exposure /ɪkˈspəʊʒə;ek-/ n. 暴露；曝光；揭露；陈列

shutter /ˈʃʌtə/ n. 快门；百叶窗；遮板 vt. 以百叶窗遮蔽

externally /eksˈtəːnəli/ adv. 外部地；外表上，外形上

trigger /ˈtrɪgə/ n. 扳机；起因；触发器 v. 触发；开动（装置）

synchronization /ˌsɪŋkrənaɪˈzeɪʃn/ n. 同步

acquisition /ˌækwɪˈzɪʃ(ə)n/ n. 获得物，获得；收购

spectral /ˈspektr(ə)l/ adj. [光] 光谱的；幽灵的；鬼怪的

sensitivity /sensɪˈtɪvɪtɪ/ n. 敏感；敏感性；过敏

filter /ˈfɪltə/ v. 过滤；渗透；用过滤法除去；缓行 n. 过滤器；筛选程序

infrared /ɪnfrəˈred/ adj. 红外线的；（设备、技术）使用红外线的

Notes and analysis

Question 1：How does the machine vision play its role in industry?

Answer：_____

Question 2：What does an industrial machine vision system involve?

Answer：_____

Question 3：List the advantages of machine vision, compared with the human eyes?

Answer：_____

Question 4：What is the basic function of the industrial camera?

Answer：_____

Question 5：How to choose a suitable camera?

Answer：_____

Question 6：List the main parameters of the industrial camera.

Answer：_____

Question 7：What is the pixel depth for a common industrial camera?

Answer：_____

Question 8：How to filter out the infrared light?

Answer：_____

Question 9：Is the industrial camera expensive?

Answer：_____

Translation

机 器 视 觉

机器视觉就是用机器代替人眼来进行测量和判断。机器视觉系统是指通过机器视觉产品将被摄取目标转换成图像信号，传送给专用的图像处理系统，根据像素分布、亮度和颜色等信息，转变为数字信号。图像系统对这些信号进行各种运算来抽取目标的特征，进而根据判别结果来控制现场的设备动作。

图 14-1 所示为一个机器视觉系统。一个典型的工业机器视觉系统由光源、镜头、相机、图像处理、处理软件、监视器、输入/输出单元等部分组成。

机器视觉相较于人类视觉有如下优点：首先，非接触式检测对观察者和被观察者不会产生模仿；其次，机器视觉系统可以长时间稳定工作；第三，机器视觉系统具有较高的测量精度；最后，拍照速度自动与被测物的速度相匹配，从而可拍摄到理想的图像。

工业相机是机器视觉系统中的一个关键组件，其最基础的功能是将光信号转变成为有序的电信号。选择合适的工业相机也是机器视觉系统设计中的重要环节，工业相机不仅直接决定了所采集到的图像分辨率、图像质量等，同时也与整个系统的运行模式直接相关。

工业相机的主要参数包括以下几种。

1. 分辨率

分辨率是指相机每次采集图像的像素点数。对于工业数字相机，分辨率一般是直接与光电传感器的像元数对应的。对于工业数字模拟相机而言，分辨率取决于视频制式。

2. 像素深度

像素深度是指即每像素数据的位数，常用的是 8Bit，工业数字相机一般还会有 10Bit、12Bit 等。

3. 最大帧率/行帧

相机采集传输图像的速率，对于面阵相机一般是每秒采集的帧数，对于线阵相机是每秒采集的行数。

4. 曝光方式和快门速度

工业线阵相机采用逐行曝光的方式，可以选择固定行频和外触发同步的采集方式，曝光时间可以与行周期一致，也可以设定一个固定的时间。面阵相机有帧曝光、场曝光和滚动行曝光等常见方式。工业数字相机一般都提供外触发彩图的功能。快门速度一般可到 $10\mu s$，高速相机可以更快。

5. 像元尺寸

像元尺寸和像元数（分辨率）共同决定了相机靶面的大小。目前，工业数字相机像元尺寸一般为 $3\sim10\mu m$，一般像元尺寸越小，制造难度越大。

6. 光谱响应特性

光谱响应特性是指该像元传感器对不同光波的敏感特性，一般响应范围为 350～1000nm。一些相机会在镜头面前增加滤镜，以滤除红外光线。

ROBOT
Chapter 15 Programmable Logic Controller

Objectives

After reading this chapter, you will be able to:

1) know the definition of programmable logic controller.

2) understand the structure of programmable logic controller.

3) be able to find the key information of programmable logic controller.

4) be familiar with functions of programmable logic controller.

5) answer the review questions at the end of the chapter.

Reading

As a typical industrial product, programmable logic controller (PLC) is a kind of digital opera-
tion electronic device designed for industrial production. PLC uses a type of programmable memory for its internal stored procedures to perform the logic operation, sequence control, timing, counting and arithmetic operations such as user-oriented instructions. It also can control various types of machinery or production process through the digital or analog input/output devices. Fig. 15-1 is a PLC.

PLC is the core part of industrial control. It is a kind of digital operation electronic system with high application rate in modern industry.

Fig. 15-1   PLC

It contains a variety of human-machine interface units and communication units, which controls the production of equipment through the input and output of digital or analog quantities. Its working principle can be divided into several steps, including input sampling, program implementation, output refresh, and then re-input sampling, implementation and output of the reciprocating work.

The PLC is essentially a computer dedicated to industrial control, and its hardware structure is basically the same as microcomputer, which mainly includes central processing unit (CPU), memory, power supply, program input device and input/output loop.

The CPU is the control center of a PLC. It receives and stores user programs and data typed from the programmer according to the functions assigned to the PLC system program. It checks the function status, such as power supply, storage, I/O, and watchdog timer. Furthermore, CPU can diagnose syntax errors in user programs.

When PLC is operating, it collects the data of each input device in the form of scanning field state and stores in the I/O image area. After that, it reads the user program in memory and carries out the command interpretation, and then the result of the logical or arithmetic operation performed according to the instructions transfers to a data register or an I/O image area. Once all the user programs are completed, the data in the output status or output register in the I/O image area is finally transferred to the corresponding output device, and the cycle runs until it stops.

The power supply of PLC is significant in the system. Without a well-performed and reliable power supply system, PLC cannot perform properly, so the design and manufacture of power supply are also very important. PLC can be connected to the AC grid directly without any treatment, when AC voltage fluctuates in the range of +10% ~ +15%.

The program input device is responsible for providing functions for the operator to input, modify, and monitor programs. The input/output loop is responsible for receiving external input element signals and external output element signals.

In the industrial production process, PLC is the center of the entire industrial control system, which is the brain of the entire workstation. The operation and stopping of a variety of mechanical equipment in the workstation are coordinated and cotrolled by PLC. In the process of changing production of the entire production line, only the PLC program needs to be changed, and the industrial robot is fine-tuning to carry out a new round of production. There are two kinds of communication transmission between industrial robot and PLC, including I/O connection and communication line connection.

Vocabulary

logic　　/'lɒdʒɪk/ n. 逻辑；逻辑学；逻辑性 adj. 逻辑的

digital　　/'dɪdʒɪt(ə)l/ n. 数字；键 adj. 数字的；手指的

procedure　　/prə'siːdʒə/ n. 程序，手续；步骤

sequence　　/'siːkw(ə)ns/ n. ［数］［计］序列；顺序；续发事件 vt. 按顺序排好

time　　/taɪm/ n. 时间；时代 adj. 定时的 vt. 计时

count　　/kaʊnt/ v. 数数；计算总数 n. 总数；数数；量的计数

arithmetic　　/ə'rɪθmətɪk/ n. 算术，算法

oriented　　/'ɔːrɪentɪd/ v. 使朝向，使面对；确定方位 adj. 以……为方向的

analog　　/'ænəlɒg/ n. ［自］模拟；类似物 adj. ［自］模拟的；有长短针的

principle　　/'prɪnsɪp(ə)l/ n. 原理，原则；主义；本质；根源

sample　　/'sɑːmp(ə)l/ v. 品尝，体验 adj. 样品的 n. 样品

refresh /rɪˈfreʃ/ vt. 更新；使……恢复 vi. 恢复精神

reciprocate /rɪˈsɪprəkeɪt/ vt. 报答；互换 vi. 往复运动；互换

alert /əˈlɜːt/ n. 警戒，警报 v. 使警觉，警告 adj. 警惕的，警觉的

diagnose /ˈdaɪəgnəʊz;-ˈnəʊz/ vt. 诊断；断定 vi. 诊断；判断

syntax /ˈsɪntæks/ n. 语法；句法；有秩序的排列

error /ˈerə/ n. 误差；错误；过失

scan /skæn/ n. 扫描；浏览 vt. 扫描；浏览 vi. 扫描；扫掠

deposit /dɪˈpɒzɪt/ n. 存款；订金；沉淀物 vt. 使沉积；存放 vi. 沉淀

pass /pɑːs/ v. 通过，经过；传递 n. 及格；经过；通行证

interpretation /ɪntɜːprɪˈteɪʃ(ə)n/ n. 解释；翻译；演出

govern /ˈgʌv(ə)n/ vt. 管理；支配；统治；控制 vi. 进行统治

provision /prəˈvɪʒ(ə)n/ n. 规定；条款；准备 vt. 供给…食物及必需品

arithmetic /əˈrɪθmətɪk/ n. 算术，算法

cycle /ˈsaɪk(ə)l/ n. 循环；周期 vt. 使循环；使轮转 vi. 循环；轮转

voltage /ˈvəʊltɪdʒ;ˈvɒltɪdʒ/ n. ［电］电压

fluctuation /ˌflʌktʃʊˈeɪʃ(ə)n;-tjʊ-/ n. 起伏，波动

grid /grɪd/ n. 网格；格子，栅格；输电网

monitor /ˈmɒnɪtə/ n. 监视器；监听器；监控器；显示屏 vt. 监控

loop /luːp/ v. 使成环；环行 n. 环状物、圈；环状结构

external /ɪkˈstɜːn(ə)l;ek-/ n. 外部；外观 adj. 外部的；表面的；外国的

signal /ˈsɪgn(ə)l/ n. 信号；暗号 adj. 显著的 vt. 标志；表示 vi. 发信号

brain /breɪn/ n. 头脑，智力；脑袋 vt. 猛击…的头部

entire /ɪnˈtaɪə;en-/ adj. 全部的，整个的；全体的

workstation n. 工作站

tune /tjuːn/ n. 曲调；和谐 vt. 调整；使一致 vi. ［电子］［通信］调谐；协调

Notes and analysis

Question 1：What is a programmable logic controller?

Answer：_____

Question 2：What can programmable logic controller be used to do?

Answer：_____

Question 3：What does the PLC hardware contain?

Answer：_____

Question 4：How does the PLC operate?

Answer：_____

Question 5：How to store the system software?

Answer：_____

64

Question 6：Why the power supply is important to a PLC controller?

Answer：_____

Question 7：What are the main kinds of communication transmission between the industrial robot and PLC?

Answer：_____

Translation

可编程逻辑控制器

作为一种常见的工业产品，可编程逻辑控制器（PLC）是专为工业生产设计的一种数字运算操作的电子装置。它采用一种可编程的存储器，用于其内部存储程序、执行逻辑运算、顺序控制、定时、计数及算术操作等面向用户的指令，并通过数字或模拟式输入/输出设备控制各种类型的机械或生产过程。图 15-1 所示为一个 PLC 控制器。

PLC 是工业控制的核心部分，是现代工业中使用率很高的一种数字运算操作电子系统。PLC 含有多种人机界面单元以及通信单元等，其通过数字量或模拟量的输入/输出以控制生产设备。PLC 工作原理可分为几个阶段，包括输入采样、程序执行和输出刷新，然后再重新进行输入采样、执行和输出。

PLC 实质是一种专用于工业控制的计算机，其硬件结构基本上与微型计算机相同，主要包括中央处理单元、存储器、电源、程序输入装置和输入/输出回路。

中央处理单元是 PLC 的控制中枢。它按照 PLC 系统程序赋予的功能接收并存储从编程器键入的用户程序和数据；检查电源、存储器、I/O 以及警戒定时器的状态，并能诊断用户程序中的语法错误。

当 PLC 投入运行时，首先它以扫描的方式接收现场各输入装置的状态和数据，并分别存入 I/O 映像区。然后从用户程序存储器中逐条读取用户程序，经过命令解释后按指令的规定执行逻辑或算数运算结果送入 I/O 映像区或数据寄存器内。所有的用户程序执行完毕之后，最后将 I/O 映像区的各输出状态或输出寄存器内的数据传送到相应的输出装置，如此循环运行，直到停止运行。

PLC 的电源在整个系统中占有重要地位。如果没有性能良好且可靠的电源系统，PLC 是无法正常工作的，因此 PLC 电源的设计和制造十分重要。一般交流电压波动在+10% ~ +15%范围内波动时，可以不采取其他措施而将 PLC 直接接入交流电网。

程序输入装置负责提供操作者输入、修改、监视程序运行的功能。输入/输出回路负责接收外部输入元件信号和外部输出元件信号。

在工业生产过程中，PLC 是整个工控系统的中枢，也就是整个工作站的大脑。协调、控制工作站中多种机械设备的运行和停止，都是由 PLC 完成的。在整个生产线的换产过程中，只需要对 PLC 程序进行更改，以及对工业机器人进行微调，即可以进行新一轮的生产。工业机器人与 PLC 之间的通信有"I/O"连接和通信线连接两种。

ROBOT
Chapter 16 Robotic Techn-ician and Engineer

Objectives

After reading this chapter, you will be able to:

1) know the job requirement of robotic engineer.
2) understand the responsibility of robotic technician.
3) be familiar with the types of technical worker.
4) know the development of engineer and technician.
5) answer the review questions at the end of the chapter.

Reading

Although the robots can replace the human worker in industrial manufacture, their work is still under the supervision by human workers, more specifically, under the managment of technicians. and engineers. Technicians (Fig. 16-1) should master the structure, installation, testing and maintenance of robots, and the maintenance is the main responsibility of robotic technician.

Fig. 16-1   A robotic technician

To make the classification in detail, technician can be classified as robot troubleshooter, robot engineer, robot repairman, automation technician and field service technician.

The duties assigned to the industrial robots include robotic drilling, robotic cutting, robotic grinding, robotic spraying, painting, robotic polishing, robotic metal forming, robotic plastics han-

Industrial Robot

dler, and robotic welding. All the previously mentioned tasks require robotic engineer who owns the academic degrees in related specialty of robotics.

The basic knowledge required for a robotic engineer includes microprocessor, programmable controller, electronic circuits, circuit analysis, mechanical sensors, and the feedback systems. Among them, the assembling issues including the disassembling and reassembling robots are fundamental. Furthermore, they are generally required to be able to repair and replace the equipment and perform preventive maintenance on robotic systems.

The professional training for technician and engineer is necessary, but quite different. For a robotic technician, a college degree is not necessary, but a lot of professional training and experience with mechanical and electrical devices are required. Compared to the technicians, engineers are required with a high-level education background, for instance, a bachelor's degree or a four-year academic degree with emphasis on mathematics, science and robotic. Engineers also hold a high-level salary which depends on work requirements and enterprise scale.

In future, an increasing demand of the trained and educated technician and engineer are received much attention due to the rapid development of industrial robot. Only the people with far-sighted work plan and actual skill/knowledge improvement can own a bright future.

Vocabulary

supervise /ˈsuːpəvaɪz;ˈsjuː-/ vt. 监督，管理；指导 vi. 监督，管理；指导

technician /tekˈnɪʃ(ə)n/ n. 技师，技术员；技巧纯熟的人

engineer /endʒɪˈnɪə/ n. 工程师 vt. 设计；策划 vi. 设计；建造

installation /ɪnstəˈleɪʃ(ə)n/ n. 安装，装置；就职

maintenance /ˈmeɪntənəns/ n. 维护，维修；保持

latter /ˈlætə/ adj. 后者的；近来的；后面的；较后的

troubleshooter /ˈtrʌblʃuːtə/ n. 解决纠纷者；故障检修工

repairer /riˈpɛərə/ n. 修理者；修补者

duty /ˈdjuːtɪ/ n. 责任；[税收] 关税；职务

assign /əˈsaɪn/ vt. 分配；指派；[计] [数] 赋值 vi. 将财产过户

disassembly /ˌdɪsəˈsemblɪ/ n. 拆卸；分解

reassembly /ˌriːəˈsemblɪ/ n. 重新聚集；重新组装；再装配

fundamental /fʌndəˈment(ə)l/ n. 基本原理；基本原则 adj. 基本的，根本的

preventive /prɪˈventɪv/ adj. 预防性的，防备的 n. 预防药，预防疗法

professional /prəˈfeʃ(ə)n(ə)l/ n. 专业人员；职业运动员 adj. 专业的；职业性的

emphasis /ˈemfəsɪs/ n. 重点；强调；加强语气

mathematics /mæθ(ə)ˈmætɪks/ n. 数学；数学运算

salary /ˈsælərɪ/ v. 给……薪金 n. 薪水，工资

wage /weɪdʒ/ v. 进行，发动；开展 n. 工资；报酬；代价

Notes and analysis

Question 1: Can an industrial robot work without human supervision? Why?

Answer: _____

Question 2: What are the main aspects that technician should understand?

Answer: _____

Question 3: List the main duties assigned to an industrial robotic technician.

Answer: _____

Question 4: Which one requires the higher education background, a technician or an engineer? Why?

Answer: _____

Question 5: What should the technician focus on with the industrial development?

Answer: _____

Translation

机器人技术人员和工程师

机器人可以代替工业生产中人类工作者的相关工作，但依然是在人类的监督下进行的，具体地说，是在技术人员和工程师的管理下进行的。技术人员应掌握机器人的结构、安装、测试和维护，其中机器人的维护是技术人员的主要职责（图 16-1）。

进一步细分，技术人员可以分为机器人故障排除员、机器人工程师、机器人修理工、自动化技术人员以及现场服务技术人员。

工业机器人可以进行钻削、切削、研磨、喷涂、喷漆、抛光、机械金属成形、机械塑料处理和焊接等工作。以上提到的所有工作都需要拥有机器人相关专业学位的机器人工程师参与。

机器人工程师需要具备的基本知识包括微处理器、可编程控制器、电子电路、电路分析、机械传感器和反馈系统。其中，装配包括机器人的拆卸和重组是机器人维护的基础问题。此外，一般要求工程师能够维修和更换设备，并对机器人系统进行预防性维护。

技术人员和工程师的专业培训是必要的，但两者之间又有很大的不同。对于机器人技术人员来说，本科学位并不是必需的，但是要求技术人员具有大量的专业培训和机电设备经验。与技术人员相比，工程师需要具有高水平的教育背景，例如，具有学士学位或四年制学位，重点学科为数学、工程和机器人。工程师的工资水平较高，一般取决于工作要求和企业规模。

随着工业机器人的快速发展，未来对训练有素的技术人员和工程技术人员的需求越来越大。技术人员只有具备有远见的工作计划和注重实际的技能/知识的提高才能拥有一个光明的未来。

Chapter 17 Robotic Research and Development

ROBOT

Objectives

After reading this chapter, you will be able to:
1) know the main research and development technology of robotics.
2) understand the development of software.
3) be able to find the key information of electrical technology.
4) classify the mechanical design technology.
5) answer the review questions at the end of the chapter.

Reading

Research and development technology of robot mainly involve the hardware structure of industrial robot control system, system structure, development of control software, robot servo communication bus technology, electrical equipment, mechanical design and so on. Fig. 17-1 shows a new designed Yumi robot.

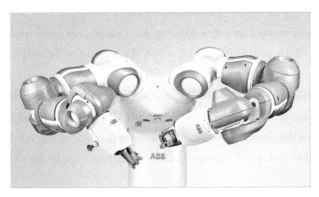

Fig. 17-1   A Yumi robot

The hardware structure of industrial robot control system mainly refers to the controller, which is the core of the robot system. With the development of microelectronics technology, cost-effective microprocessor brings new development opportunities for robot controller, which makes it possible to develop the robot controller with low-cost and high-performance. In order to ensure sufficient computing and storage capacity of the system, most of the robot controllers are composed of ARM series, DSP series, POWERPC series, Intel series and other chips with strong computing capacities. In addition, because the existing general chips on the functionality and performance cannot completely satisfy certain robot system in terms of price, performance, integration and interface requirements, it creates a demand for embedded System on Chip (SoC) technology for robotic systems. Integrating a specific processor with the required interface, which can simplify the design of peripheral circuit to reduce system size and cost.

In the aspect of controller architecture, the research focuses on the division of functions and the specification of information exchange between functions. In the research of open controller architecture, there are two basic structures. One is a simple structure based on hardware level division. The other one is the structure based on function division, which considers the hardware and software together, and also it is the direction of research and development of robot controller architecture.

In terms of robot software development environment, general industrial robot companies have their own independent development environment and independent robot programming language. At present, a lot of research work provides amounts of open source code, which can be integrated in the part of robot hardware structure and control the operation. From the perspective of the development of the robot industry, there are two requirements for the robot software development environment. One aspect is from the end users of robots. They not only use robots, but also set functions to robots through programming, which is often implemented in visual programming language.

At present, there is no servo communication bus specially used in the robot system in the world. In the practical application process, some commonly used buses, such as Ethernet, CAN, 1394, SERCOS, USB, RS-485 and so on, are usually used in the robot system according to the system requirements. Currently, most communication control buses can be classified into two categories, including the serial bus technology based on RS-485 and wire-driven technology and high-speed serial bus technology based on real-time industrial Ethernet.

In terms of electrical technology, it mainly includes pneumatic, electrical control and PLC programming technology, which can prepare and adjust the robot station control program according to the requirements of the production line.

In terms of mechanical design technology, it is necessary to master the drawing design technology, like mechanical drawing CAD and electronic circuit CAD. Furthermore, it is essential to know the structural installation diagram and electrical schematic diagram of the robot application system. It is able to assembly molds and apply standard parts, design and apply the engineering examples.

Vocabulary

hardware   /'hɑːdweə/ n. 计算机硬件；五金器具

communication   /kəmjuːnɪ'keɪʃ(ə)n/ n. 通讯，[通信] 通信；交流；信函

technology   /tek'nɒlədʒɪ/ n. 技术；工艺；术语

controller   /kən'trəulə/ n. 控制器；管理员；主计长

microelectronic　/ˌmaɪkrəʊɪˌlek'trɒnɪk/ *adj.* ［电子］微电子的

opportunity　/ˌɒpə'tjuːnəti/ *n.* 时机，机会

ensure　/ɪn'ʃɔː;-'ʃʊə;en-/ *vt.* 保证，确保；使安全

sufficient　/sə'fɪʃ(ə)nt/ *adj.* 足够的；充分的

storage　/'stɔːrɪdʒ/ *n.* 存储；仓库；贮藏所

series　/'sɪəriːz;-rɪz/ *n.* 系列，连续；［电］串联；级数；丛书

chip　/tʃɪp/ *n.* 芯片，晶片；碎片 *v.* 打缺；铲，凿，削

existing　/ɪg'zɪstɪŋ/ *v.* 存在；被发现 *adj.* 存在的；现行的

genera　/'dʒenərə/ *n.* ［生物］属（genus 的复数形式）；种；类

functionality　/fʌŋkʃə'næləti/ *n.* 功能；［数］泛函性，函数性

satisfy　/'sætɪsfaɪ/ *vi.* 令人满意；令人满足 *vt.* 满足；说服；使满意

price　/praɪs/ *n.* 价格；价值；代价 *vt.* 给……定价；问……的价格

integration　/ɪntɪ'greɪʃ(ə)n/ *n.* 集成；综合

peripheral　/pə'rɪf(ə)r(ə)l/ *adj.* 外围的；次要的 *n.* 外部设备

circuit　/'sɜːkɪt/ *n.* ［电子］电路，回路 *vt.* 绕回…环行 *vi.* 环行

aspect　/'æspekt/ *n.* 方面；方向；形势；外貌

architecture　/'ɑːkɪtektʃə/ *n.* 建筑学；建筑风格；建筑式样；架构

focus　/'fəʊkəs/ *n.* 焦点；中心；焦距 *vt.* 使集中 *vi.* 集中；调节焦距

exchange　/ɪks'tʃeɪndʒ/ *n.* 交换；交流；兑换 *vt./vi.* 交换；交易；兑换

integrate　/'ɪntɪgreɪt/ *n.* 一体化 *adj.* 整合的 *vi.* 求积分 *vt.* 使…完整

perspective　/pə'spektɪv/ *n.* 观点；远景；透视图 *adj.* 透视的

implement　/'ɪmplɪm(ə)nt/ *n.* 工具；手段 *vt.* 实施，执行；实现

visual　/'vɪʒjʊəl;-zj-/ *adj.* 视觉的，视力的

real-time　/ˌrɪəl'taɪm/ *adj.* 实时的；接到指示立即执行的

adjustment　/ə'dʒʌs(t)m(ə)nt/ *n.* 调整，调节；调节器

draw　/drɔː/ *n.* 平局；抽签 *vi.* 拉；拖 *vt.* 画；拉；吸引

schematic　/skiː'mætɪk;skɪ-/ *n.* 原理图；图解视图 *adj.* 图解的；概要的

Notes and analysis

Question 1：Is it necessary to research and develop the technology of industrial robot? Why?

Answer：_____

Question 2：What does the hardware structure include?

Answer：_____

Question 3：List the commonly used servo communication buses.

Answer：_____

Question 4：List the main electrical technologies.

Answer：_____

Question 5：List the main mechanical design technologies.

Answer: _____

Translation

机器人研发技术

工业机器人研发技术主要涉及工业机器人控制系统硬件结构、体系结构、控制软件开发、机器人伺服通信总线技术、电气设备和机械设计等。图 17-1 所示为新型 Yumi 机器人。

工业机器人控制系统硬件结构主要指控制器，它是机器人系统的核心。随着微电子技术的发展，高性价比的微处理器为机器人控制器带来了新的发展机遇，使开发低成本、高性能的机器人控制器成为可能。为了保证系统具有足够的计算与存储能力，目前机器人控制器多采用计算能力较强的 ARM 系列、DSP 系列、POWERPC 系列和 Intel 系列等芯片。此外，由于已有的通用芯片在功能和性能上不能完全满足某些机器人系统在价格、性能、集成度和接口等方面的要求，这就产生了机器人系统对嵌入式系统技术的需求。将特定的处理器与所需要的接口集成在一起，可简化系统外围电路的设计，缩小系统尺寸，并降低成本。

控制器体系结构方面，其研究重点是功能划分和功能之间信息交换的规范。在开放式控制器体系结构研究方面，有两种基本结构，一种是基于硬件层次划分的结构，该类型结构比较简单；另一种是基于功能划分的结构，它将软硬件一同考虑，它也是机器人控制器体系结构研究和发展的方向。

在机器人软件开发环境方面，一般工业机器人公司都有自己独立的开发环境和独立的机器人编程语言。目前在机器人开发环境方面已有大量研究工作，提供了很多开放源码，可在部分机器人硬件结构下进行集成和控制操作。从机器人产业发展来看，对机器人软件开发环境有两方面的需求。其中一方面是来自机器人的最终用户，他们不仅使用机器人，而且希望能够通过编程的方式赋予机器人更多的功能，这种编程往往是采用可视化编程语言实现的。

目前国际上还没有专用于机器人系统的伺服通信总线。在实际应用过程中，通常根据系统需求，把常用的一些总线，如以太网、CAN、1394、SERCOS、USB、RS-485 等用于机器人系统中。当前大部分通信控制总线可以归纳为两类，即基于 RS-485 和线驱动技术的串行总线技术和基于实时工业以太网的高速串行总线技术。

电气技术主要涉及技术包括气动、电气控制与 PLC 编程技术，能根据生产线的工序要求，编制、调整机器人工作站控制程序。

关于机械设计技术，要掌握机械制图 CAD、电子线路 CAD 等绘图设计技术，能读懂机器人应用系统的结构安装图和电气原理图，会装配模具，运用标准件，会设计和运用工程实例等。

Chapter 18 Robot Manual

Objectives

After reading this chapter, you will be able to:
1) know the main parts of a manual.
2) understand the contents of a manual.
3) can be instructed under the guide of a manual.
4) be able to find the key information of a manual.
5) answer the review questions at the end of the chapter.

Reading

Product manual, as shown in Fig. 18-1, refers to a relatively detailed description of the product in text, with the purpose of enabling users to recognize the product, use the product, and maintain the product. Thus, a manual must not exaggerate the role and performance of the product at will. Product manual must be authentic, scientific, clearly structured, and practical.

Industrial product manual mainly includes title cover, catalog, and main content. The title is generally named as "product name manu-

Fig. 18-1　A manual book

al", which is located on the cover page of the manual. The name and LOGO of the product supplier are also on the cover page which focuses on the visual effect and specific page design. The catalog provides an index based on the body content. The content text is the main body of product manual, introducing the product information, structure, characteristics, performance, usage, maintenance, matters attention, contact information and so on.

Standard product specifications must include the following issues.

1. Authenticity

For products, especially industrial products, the direct cost and indirect cost are both high, which is directly related to the interests of buyers and consumers. Therefore, a manual must assure the authenticity of content, and it should not contain the false information and unreal content.

2. Organization

The product specification requires a reasonable structure and a clear logical introduction. When using a product manual, it is generally based on the comprehensive introduction to the partial introduction, from the easy part to the hard content. In addition, the instruction cases are compulsory for the customers.

3. Popularity

Product manual should use simple language to describe and guide based on the guidance principle. The aim of a manual is to make the customer understand the product clearly.

In addition, the product manual commonly includes case description and breakdown maintenance knowledge, so that users can query professional information when they encounter problems. Meanwhile, in order to make a better illustration, a large number of pictures, charts and graphs are adopted in the industrial product manual. With the change in the times, the electronic animation and guidance are also included as the supporting file, which aims to provide better services to the customers.

Vocabulary

manual /ˈmænjʊ(ə)l/ n. 说明书；小册子 adj. 体力的；手控的

relatively /ˈrelətɪvlɪ/ adv. 相当地；相对地，比较地

detail /ˈdiːteɪl/ n. 细节，琐事；具体信息 vt. 详述；选派 vi. 画详图

description /dɪˈskrɪpʃ(ə)n/ n. 描述，描写；类型；说明书

text /tekst/ n. [计] 文本；课文；主题 vt. 发短信

enable /ɪnˈeɪb(ə)l;en-/ v. 使能够；使成为可能；（计算机）启动

exaggerate /ɪgˈzædʒəreɪt;eg-/ vt. 使扩大；使增大 vi. 夸大；夸张

authentic /ɔːˈθentɪk/ adj. 真正的，真实的；可信的

scientific /saɪənˈtɪfɪk/ adj. 科学的，系统的

practical /ˈpræktɪk(ə)l/ adj. 实际的；实用性的

title /ˈtaɪt(ə)l/ n. 标题；头衔；权利 adj. 标题的 vt. 加标题于

catalog /ˈkætəlɒg/ n. [图情] [计] 目录；登记 vt. 登记；为…编目录 vi. 编目录

content /kənˈtent/ n. 内容，目录；容量 adj. 满意的 vt. 使满足

supplier /səˈplaɪə/ n. 供应厂商，供应国；供应者

consumer /kənˈsjuːmə/ n. 消费者；用户，顾客

false /fɔːls;fɒls/ adj. 错误的；虚伪的；伪造的 adv. 欺诈地

organization /ˌɔːgənaɪˈzeɪʃn;ˌɔː/ n. 组织；机构；体制；团体

comprehensive /ˌkɒmprɪˈhensɪv/ n. 综合学校 adj. 综合的；广泛的

local /ˈləʊk(ə)l/ adj. 当地的，局部的；局域的 n. 当地人；局部

shallow /ˈʃæləʊ/ n. [地理] 浅滩 adj. 浅的；肤浅的 vt. 使变浅 vi. 变浅

deep /diːp/ adj. 深的 adv. 在深处，深深地；边线地

assist /əˈsɪst/ v. 参加，出席；有助益 n. 帮助；助攻

eliminate　/ɪˈlɪmɪneɪt/ *vt.* 消除；排除

obscure　/əbˈskjʊə/ *adj.* 不清楚的；隐蔽的 *vt.* 掩盖；隐藏

query　/ˈkwɪərɪ/ *n.* 疑问；疑问号；[计] 查询 *vi.* 询问 *vt.* 询问

encounter　/ɪnˈkaʊntə;en-/ *v.* 遭遇；邂逅；遇到 *n.* 遭遇；偶然碰见

reflect　/rɪˈflekt/ *vt.* 反映；反射；表达；显示 *vi.* 反射，映现；深思

illustration　/ɪləˈstreɪʃ(ə)n/ *n.* 说明；插图；例证；图解

picture　/ˈpɪktʃə/ *n.* 照片；影片；景色 *vt.* 画；想象；描写

chart　/tʃɑːt/ *n.* 图表；图纸；排行榜 *vt.* 绘制…的图表；记录

graph　/grɑːf/ *n.* 图表；曲线图 *vt.* 用曲线图表示

animation　/ænɪˈmeɪʃ(ə)n/ *n.* 活泼；激励；动画

Notes and analysis

Question 1：What is a manual?

Answer：_____

Question 2：What is the standard structure of a manual?

Answer：_____

Question 3：How to prepare a high-quality manual?

Answer：_____

Question 4：Is there any image contained in the manual?

Answer：_____

Translation

机器人说明书

产品说明书（图 18-1）是指用文本对产品进行的详细表述，目的是使用户能够认识产品、使用产品以及维护产品。制作产品说明书时，不可随意夸大产品的作用和性能。产品说明书必须具有真实性、科学性，结构清晰，有实用指导作用。

工业类产品说明书主要包括标题封面、目录和正文。标题一般为"产品名称说明书"，位于说明书封面页。封面一般也包括产品提供方名称以及企业 LOGO，注重视觉效果，设计多样。目录为正文内容提供索引。正文是产品说明书的主体，介绍产品信息、结构、特征、性能、使用方法、保养维护、注意事项和联系方式等内容。

标准的产品说明书必须具备以下几点。

1. 真实性

对于产品，尤其是工业产品，产品的直接费用和产生的间接费用均较高，直接关系到买方和消费者的利益。因此，必须要保证产品说明书内容的真实性，切勿添加虚假或者不符合实际的内容。

2. 条理性

产品说明书要求有合理的结构和清晰的逻辑介绍。使用产品说明书时一般会根据由综合介绍到局部介绍的原则，由浅入深，同时有案例辅助。

3. 通俗性

产品说明书应该使用通俗简单的语言，以指导性为原则，清晰明了地介绍产品。

此外，一般常用的产品说明书都包括案例使用介绍以及故障维护知识，便于使用者遇到问题时能查询到专业信息。同时，为了更好地体现说明效果，工业产品说明书会使用大量的图片、图表。根据时代变化，目前有一些工业设备除了纸质产品说明书外，还附加电子动画类说明书，从而更好地指导客户使用。

Chapter**19** Safety Considerations

Objectives

After reading this chapter, you will be able to:

1) know the common safety rules.

2) have the production safety consciousness.

3) answer the review questions at the end of the chapter.

Reading

Safety issue (Fig. 19-1) in the robot working area mainly includes teaching and production.

1. Teaching and manual robot

1) Robot operations must be operated under the guidance of personnel who have received systematic training or process control.

2) The robot working area must be clean.

3) Do not wear gloves during the operation.

4) Ensure that the control power supply is in good condition.

Fig. 19-1   Work safety

5) Technical workers are not allowed to stay in the robot's activity range (within the safety fence) before starting the machine.

6) Know the movement trend of robot before its start.

7) Confirm the speed of the robot's movements.

8) Once danger is foreseen, the emergency stop switch should be quickly pressed to stop the robot.

9) Before leaving the working area of the equipment for any reason, the emergency stop switch should be pressed to avoid sudden power failure or shutdown loss of zero position, and the instructor should be placed in a safe position.

2. Production operation

1) Operator must understand all the tasks of the robot before starting up and running.

2）Ensure that the robot is at the origin point or designated position.

3）Ensure that the robot loads the program correctly.

4）No worker is in the robot's activity area（within the safety fence）.

5）Before the program runs, it is compulsory to check the whole system status, confirm no remote instructions to the peripheral equipment, and clean the potential threat to the user.

6）Confirm the moving switch, sensor and control signal maintenance position and status of the robot.

7）Confirm the position of the robot's human controller and the emergency button on the containment control device and operate it accordingly.

8）Stop the robot halfway does not mean that the program has finished. Do not enter the working area of the robot in a static state.

9）Once the failure occurs, it must carry out the strict inspection. After the maintenance of fault elimination is correct, the robot must run at the low speed before the automatic operation.

10）It is required to strictly obey the routine maintenance regulations.

During the operations, the basic safety symbols should be understood. There are dozens of safety symbols, and some of them are displayed in Fig. 19-2.

Fig. 19-2　Safety symbols

Vocabulary

safety　/'seifti/ *n.* 安全；保险；安全设备

issue　/'ifuː；'ısjuː/ *n.* 问题；发行物 *vt.* 发行，发布 *vi.* 发行；流出

operation　/ɒpəˈreɪʃ(ə)n/ *n.* 操作；经营；[外科] 手术；[数] [计] 运算

guide　/gaɪd/ *n.* 指南；向导；入门书 *vt.* 引导；带领；操纵 *vi.* 担任向导

systematic　/sɪstəˈmætɪk/ *adj.* 系统的；体系的；[图情] 分类的；一贯的

wear　/weə/ *v.* 穿，戴；耐用 *n.* 衣物；磨损；耐久性

personnel /pɜːsəˈnel/ *n.* 人事部门；全体人员 *adj.* 人员的；有关人事的

activity /ækˈtɪvɪtɪ/ *n.* 活动；行动；活跃

foresee /fɔːˈsiː/ *vt.* 预见；预知

emergency /ɪˈmɜːdʒ(ə)nsɪ/ *n.* 紧急情况；突发事件 *adj.* 紧急的；备用的

switch /swɪtʃ/ *n.* 开关；转变 *v.* 改变（立场、方向等）；替换；调换

shutdown /ˈʃʌtdaʊn/ *n.* 关机；停工；关门；停播

instructor /ɪnˈstrʌktə/ *n.* 指导书；教员；指导者

assurance /əˈʃʊər(ə)ns/ *n.* 保证，担保；（人寿）保险；确信

remote /rɪˈməʊt/ *n.* 远程 *adj.* 遥远的；偏僻的；疏远的

peripheral /pəˈrɪf(ə)r(ə)l/ *adj.* 外围的；次要的 *n.* 外部设备

device /dɪˈvaɪs/ *n.* 装置；策略；图案；设备；终端

threat /θret/ *n.* 威胁，恐吓；凶兆

containment /kənˈteɪnm(ə)nt/ *n.* 控制，抑制；遏制；封锁（政策）

halfway /ˈhɑːfweɪ/ *adj.* 中途的；不彻底的 *adv.* 到一半；在中途

static /ˈstætɪk/ *n.* 静电；静电干扰 *adj.* 静态的；静电的；静力的

failure /ˈfeɪljə/ *n.* 失败；故障；失败者；破产

strict /strɪkt/ *adj.* 严格的；绝对的；精确的；详细的

fault /fɔːlt/ *n.* 故障；[地质] 断层；错误 *vi.* 弄错；产生断层

elimination /ɪˌlɪmɪˈneɪʃən/ *n.* 消除；淘汰；除去

Notes and analysis

True or false：

1）The working area should be clean.   (   )

2）Wearing gloves to operate the teaching tray and operating tray is allowed.   (   )

3）The control power is kept at the maximum speed.   (   )

4）There is no need to check the speed before robot moves.   (   )

5）Once danger is foreseen, the operator should run quickly without pressing the emergency stop switch.   (   )

6）Assure the program is executable and safe.   (   )

7）Technical workers can stay in the robot's activity area.   (   )

8）Before the program runs, there is no need to check the whole system status.   (   )

9）The locations of robot's controller and emergency switch can be anywhere.   (   )

10）After the failure, it must carry out the strict robot inspection.   (   )

Translation

安 全 事 项

机器人使用安全事项主要包括示教安全事项和生产运行安全事项（图 19-1）。

1. 示教和手动机器人

1）机器人的操作必须由接受过系统培训的人员或在掌握操作流程的人员指导下操作。

2）机器人周围区域必须清洁整齐。

3）不允许戴手套操作示教器。

4）确保控制电源完好。

5）开机之前确认机器人活动范围（安全护栏以内）内无任何人员。

6）启动之前要考虑到机器人的运动趋势。

7）确认机器人的运行速度。

8）一旦预见要发生危险，应迅速按下急停开关使机器人停止运动。

9）因故离开设备工作区域前应按下急停开关，避免突然断电或者关机零位丢失，并将示教器放置在安全位置。

2. 生产运行

1）在开机运行前，确保操作者了解机器人要执行的全部任务。

2）确保机器人在原点或者指定位置。

3）确保机器人正确加载程序。

4）确认机器人活动范围（安全护栏以内）内无任何人员。

5）程序运行前，检查整个系统状态，确认没有到外围设备的远程指令，确认没有动作对使用者有威胁。

6）确认机器人移动的开关、传感器和控制信号处于保养位置和状态。

7）确认机器人控制器和外围控制设备上的紧急停止按钮的位置并会操作。

8）机器人中途停止不代表程序运行完毕，切勿随意进入静止状态机器人工作区域。

9）发生故障后，必须进行严格检查，故障排除无误后，须低速试运行无误后方可自动运行。

10）严格遵守机器的日常维护工作制度。

在操作过程中，应掌握并注意安全标志。安全标志数量较多，图 19-2 所示为其中一部分。

Chapter 20 Robot Maintenance

Objectives

After reading this chapter, you will be able to:

1) know the definition of maintenance.
2) understand the role of maintenance.
3) be familiar with maintenance issues.
4) be able to maintain industrial robots.
5) answer the review questions at the end of the chapter.

Reading

The management and maintenance of industrial robot is an essential part in the process of using industrial robot. The purpose of maintenance is to reduce the failure rate and downtime of the robot, and further maximize production efficiency.

The maintenance of industrial robot mainly aims at the control cabinet and robot body (Fig. 20-1). According to the maintenance characteristics, it mainly includes general maintenance and routine maintenance. General maintenance means that the operator of the industrial robot cheeks the equipment before starting up to confirm the integrity of equipment and origin position of the robot. The operator should pay attention to the robot operation in the working process, including oil level, gauge pressure, indication signal, safety device, etc. Routine maintenance is divided into control cabinet maintenance and robot body system maintenance.

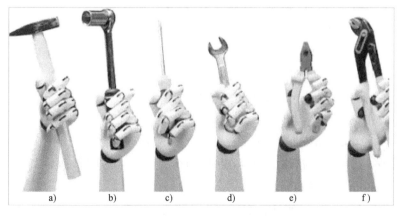

Fig. 20-1 Tools for robot maintenance

For the maintenance of robot control cabinet, it generally includes daily cleaning and mainte-nance, measuring system battery, checking fan unit and cooler, etc. The maintenance interval mainly depends on the environmental conditions, the operating hours and temperature of the robot.

For the maintenance of industrial robot body, it mainly involves cleaning and inspection of the industrial manipulator, lubrication of the reducer, and the shaft braking test of the mechanical arm. Industrial manipulator bases and arms always need to be cleaned regularly to avoid accumulation of dust and particles. The inspection of the industrial manipulator includes checking whether the bolts are loose or slipping. It is necessary to check whether loose part is normal, and the speed change is complete, whether operating system is under safety protection, and safety device is flexible and reli-able. Also, the equipment must be in good condition without corrosion, knocking down, pulling a-way, oil leakage, water, electricity and other phenomenon. The surrounding ground is clean and ti-dy, without oil, debris and so on. For the lubrication issues, the focuses are on the lubrication con-dition and regular quantity of lubricating oil.

The shaft brake test is to determine whether the brake is working normally, as each shaft motor brake shall wear in the process of operation. Repeated testing is required during the life of the ma-chine to verify that the machine maintains its original capabilities.

Vocabulary

cabinet /ˈkæbɪnɪt/ n. 内阁；橱柜 adj. 内阁的；私下的

routine /ruːˈtiːn/ n. 常规，惯例 adj. 常规的，例行的 v. 按惯例安排

operator /ˈɒpəreɪtə/ n. 经营者；操作员；运营商

spot /spɒt/ n. 地点；斑点 adj. 现场的 vt. 认出

inspection /ɪnˈspekʃn/ n. 视察，检查

integrity /ɪnˈtegrɪtɪ/ n. 完整；正直；诚实；廉正

origin /ˈɒrɪdʒɪn/ n. 起源；原点；出身；开端

oil /ɒɪl/ n. 油；石油 vt. 加油；使融化 vi. 融化；加燃油

gauge /geɪdʒ/ n. 计量器；标准尺寸；容量规格 vt. 测量；估计；给…定规格

battery /ˈbætri/ n. [电]电池，蓄电池

interval /ˈɪntəv(ə)l/ n. 间隔；间距；幕间休息

lubrication /ˌluːbrɪˈkeɪʃən/ n. 润滑；润滑作用

brake /breɪk/ n. 刹车；阻碍

accumulation /əkjuːmjʊˈleɪʃ(ə)n/ n. 积聚，累积；堆积物

loose /luːs/ n. 放纵 adj. 宽松的；不牢固的 vt. 释放 adv. 松散地

slipping /ˈslɪpɪŋ/ n. 滑动 v. 滑动 adj. 渐渐松弛的

corrosion /kəˈrəʊʒn/ n. 腐蚀；腐蚀产生的物质；衰败

phenomena /fəˈnɒmɪnə/ n. 现象

tidy　　/'taɪdɪ/ *adj.* 整齐的 *vi.* 整理；收拾 *vt.* 整理；收拾

debris　　/'deɪbriː/ *n.* 碎片，残骸

original　　/ə'rɪdʒɪn(ə)l/ *adj.* 原来的；开始的；创新的；原作的 *n.* 原件

capability　　/ˌkeɪpə'bɪləti/ *n.* 才能，能力；性能，容量

Notes and analysis

Question 1：What is the purpose of maintenance?

Answer：_____

Question 2：What does the maintenance focus on?

Answer：_____

Question 3：What is the aim of general maintenance?

Answer：_____

Question 4：What does the routine maintenance contain?

Answer：_____

Question 5：How to carry out the maintenance of robot control cabinet?

Answer：_____

Question 6：What are the necessary issues to maintain the industrial robot body?

Answer：_____

Question 7：What is a reasonable maintenance frequency for an industrial robot?

Answer：_____

Translation

机器人的维护保养

工业机器人的管理与维护保养是工业机器人使用过程中必不可少的部分。维护保养的目的是减少机器人的故障率和停机时间，最大限度地提高生产效率。

工业机器人的维护保养主要针对控制柜和机器人本体，图 20-1 所示为机器人维护保养工具。根据维护保养的特点，通常有一般性保养和例行维护之分。一般性保养是指工业机器人操作者在开机前对设备进行点检，确认设备完好以及机器人的原点位置。在工作过程中，注意机器人的运行情况，包括油标、油位、仪表压力、指示信号和保险装置等。例行维护分为控制柜维护和机器人本体系统的维护。

机器人控制柜的维护保养一般包括日常清洁维护、测量系统电池、检查风扇单元以及冷却器等。保养时间间隔主要取决于环境条件，机器人运行时数以及温度。

工业机器人本体维护主要包括工业机械手的清洗和检查、减速器的润滑以及机械手的轴制动测试。工业机械手底座和手臂需要定期清洗，以避免灰尘和颗粒物堆积。工业机械手的检查包括检查各螺栓是否有松动、滑丝现象；易松劲脱离部位是否正常；变速是否齐全，操作系统安全保护、保险装置等是否灵活可靠；检查设备有无腐蚀、碰砸、拉离和漏油、水、

电等现象；周围地面应清洁、整齐，无油污、杂物等；检查润滑情况，并定时定点加入定质定量的润滑油。

轴制动测试是为了确定制动器是否正常工作，因为在操作过程中，每个轴电动机制动器都会正常磨损，必须进行测试。在机器使用寿命期间需要反复测试，以确保机器能维持原来的功能。

Chapter**21** Industrial Robot and Industry 4.0

Objectives

After reading this chapter, you will be able to:
1) understand the content of Industry 4.0.
2) know the history of industry development.
3) certify the characteristics of Industry 4.0.
4) point out the differences between industry eras.
5) answer the review questions at the end of the chapter.

Reading

Industry 4.0 is an artificial definition according to different stages of industrial development, as shown in Fig. 21-1. Currently, it is believed that industry 1.0 represents the era of steam engine. Industry 2.0 and industry 3.0 represent the era of electrification and the era of information technology, respectively. Industry 4.0 is the era of using information technology to promote industrial transformation (Fig. 21-2), which is the era of intelligent.

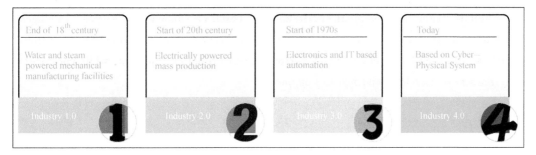

End of 18th century	Start of 20th century	Start of 1970s	Today
Water and steam powered mechanical manufacturing facilities	Electrically powered mass production	Electronics and IT based automation	Based on Cyber-Physical System
Industry 1.0 **1**	Industry 2.0 **2**	Industry 3.0 **3**	Industry 4.0 **4**

Fig. 21-1　Industry development

In 2013, the concept of industry 4.0 was officially launched at the Hannover Messe in Germany, which aimed to improve the competitiveness of German industry and seize the first opportunity in the new round of industrial revolution. According to German academia and industry, the concept of "industry 4.0" is the fourth industrial revolution or revolutionary production method led by intelli-

Fig. 21-2　Industry 4. 0

gent manufacturing. The strategy aims to transform the manufacturing industry to intelligentization by making full use of the combination of information and communication technology and cyberspace virtual system - information physical system.

Industry 4. 0 drives a new round of industrial revolution, with the internet as its core feature. Internet technology has reduced the information asymmetry between production and marketing, and it has accelerated the interaction and feedback between them. Thus, it creates a consumer-driven business model, and industry 4. 0 is the key link to achieve this model. Industry 4. 0 represents the intelligent production of "Internet + Manufacturing" and nurtures a large number of new business models.

Industry 4. 0 can be divided into intelligent factories, intelligent production and intelligent logistics. Intelligent factory focuses on research of intelligent production systems and processes, as well as the realization of networked distributed production facilities. Intelligent production mainly involves production logistics management, man-machine interaction and the application of 3D technology in industrial production processes. The plan will focus on attracting small and medium-size enterprise to become users and beneficiaries of a new generation of intelligent production technologies, as well as creators and suppliers of advanced industrial production technologies. Intelligent logistics mainly integrate logistics resources through the internet, Internet of Things and logistics networks to give full play to the efficiency of existing logistics resource suppliers, while consumers can quickly get service matching and logistics support.

Vocabulary

artificial　/ˌɑːtɪˈfɪʃl/ *adj.* 人造的；仿造的；非原产地的；武断的

era　/ˈɪərə/ *n.* 时代；年代；纪元

steam　/stiːm/ *n.* 蒸汽；蒸汽动力 *vi.* 冒水汽 *adj.* 蒸汽的

electrification　/iˌlektrɪfɪˈkeɪʃən/ *n.* 电气化；带电；充电

information　/ɪnfəˈmeɪʃ(ə)n/ *n.* 信息，资料；知识；情报

intelligent　/ɪnˈtelɪdʒ(ə)nt/ *adj.* 智能的；聪明的；理解力强的

concept   /ˈkɒnsept/ n. 观念，概念

launch   /lɔːntʃ/ v. 发射，发动 n. 发射；发行

exposition   /ekspəˈzɪʃ(ə)n/ n. 博览会；阐述；展览会

competitiveness   /kəmˈpetətɪvnɪs/ n. 竞争力

seize   /siːz/ vt. 抓住；夺取；逮捕 vi. 抓住；利用

opportunity   /ˌɒpəˈtjuːnəti/ n. 时机，机会

revolution   /revəˈluːʃ(ə)n/ n. 革命；旋转；运行；循环

essence   /ˈes(ə)ns/ n. 本质，实质；精华

scope   /skəʊp/ n. 范围；余地；视野；眼界 vt. 审视

customize   /ˈkʌstəmaɪz/ vt. 定做，按客户具体要求制造

reform   /rɪˈfɔːm/ v. 改革；重组 n. 改革，改良；改正 adj. 改革的

crucial   /ˈkruːʃ(ə)l/ adj. 重要的；决定性的；定局的；决断的

round   /raʊnd/ adj. 圆的；弧形的 n. 阶段；轮次 prep. 围绕；绕过

asymmetry   /eɪˈsɪmɪtrɪ/ n. 不对称

feedback   /ˈfiːdbæk/ n. 反馈；资料；回复

link   /lɪŋk/ n. [计] 环节；联系 vt. 连接，联结 vi. 连接起来

achieve   /əˈtʃiːv/ vt. 取得；成功 vi. 达到预期的目的

academia   /ˌækəˈdiːmɪə/ n. 学术界；学术生涯

distribute   /dɪˈstrɪbjuːt/ vt. 分配；散布；分开；把…分类

involve   /ɪnˈvɒlv/ vt. 包含；牵涉；使陷于；潜心于

enterprise   /ˈentəpraɪz/ n. 企业；事业；进取心；事业心

beneficiary   /benɪˈfɪʃ(ə)rɪ/ n. [金融] 受益人，受惠者 adj. 拥有封地的

generation   /dʒenəˈreɪʃ(ə)n/ n. 一代；产生；一代人；生殖

creator   /kriːˈeɪtə/ n. 创造者；创建者

Notes and analysis

Question 1：What are the main developments of the industry?

Answer：_____

Question 2：When and where the Industry 4.0 was named?

Answer：_____

Question 3：Why does German focus on Industry 4.0?

Answer：_____

Question 4：What are the main contents of Industry 4.0?

Answer：_____

Question 5：Is Industry 4.0 important to China? Why?

Answer：_____

Translation

工业机器人与工业 4.0

工业 4.0 是根据工业发展的不同阶段而进行的人为划分，如图 21-1 所示。目前认为，工业 1.0 是蒸汽机时代，工业 2.0 是电气化时代，工业 3.0 是信息化时代，工业 4.0 则是利用信息化技术促进产业变革的时代（图 21-2），也就是智能化时代。

2013 年，德国汉诺威工业博览会上正式推出工业 4.0 的概念，其目的是为了提高德国工业的竞争力，在新一轮工业革命中占领先机。德国学术界和产业界认为，"工业 4.0" 概念是以智能制造为主导的第四次工业革命或革命性的生产方法。该战略旨在通过充分利用信息通信技术和网络空间虚拟系统—信息物理系统相结合的手段，引领制造业向智能化转型。

工业 4.0 驱动新一轮工业革命，其核心特征是互联网。互联网技术降低了产销之间的信息不对称，加速了两者之间的相互联系和反馈，因此，催生出消费者驱动的商业模式，而工业 4.0 是实现这一模式关键环节。工业 4.0 代表了 "互联网+制造业" 的智能生产，孕育着大量的新型商业模式。

"工业 4.0" 具体可以分为智能工厂、智能生产以及智能物流。智能工厂重点研究智能化生产系统及过程，以及网络化分布式生产设施的实现。智能生产主要涉及整个企业的生产物流管理、人机互动以及 3D 技术在工业生产过程中的应用等。该计划将特别注重吸引中小企业参与，力图使中小企业成为新一代智能化生产技术的使用者和受益者，同时也成为先进工业生产技术的创造者和供应者。智能物流主要通过互联网、物联网、物流网整合物流资源，充分发挥现有物流资源供应方的效率，而需求方则能够快速获得服务匹配，得到物流支持。

Chapter 22 Robot Policy

Objectives

After reading this chapter, you will be able to:

1) know the robot policy in China.

2) understand the aim of robot policy.

3) be able to find the connection between the policy and industrial robot.

4) answer the review questions at the end of the chapter.

Reading

In the comprehensive considerations of future international development trends and industrial development of China's realistic foundation conditions, based on the overall requirements of China industrialization and accelerating transformation of economic development mode, much attention has been put to speed up the implementation from quantity scale industrial power to the quality scale industrial power. As shown in Fig. 22-1, China aims to transform Made in China to Created in China, Speed of China to Quality of China, and Product of China to Brand of China, so as to basically achieve industrialization and become a manufacturing power.

Fig. 22-1   Keyword of China Policy

Currently, China government initiatively promotes the development of ten key areas, namely information technology, numerical control tools and robots, aerospace equipment, marine engineering equipment and high-tech ships, railway equipment, energy-saving new energy vehicles, electric equipment, new materials, biomedicine and medical instruments, and agricultural machinery.

The development direction of industrial robot includes two parts. Firstly, it is to develop the product series of industrial robot bodies and key components, promote the industrialization and application of industrial robots, and meet the urgent needs of transformation and upgrading of manufacturing industry. Secondly, it is to make breakthrough in key technologies of intelligent robot, develop a number of intelligent robot, and cope with the challenge of a new round of technological revolution and industrial transformation.

China sold 135, 000 units of industrial robots in 2018, which is about 1/3 of global sales. Chinese robot manufacturers are expanding the share of the domestic market. From the perspective of the industrial chain, enterprises located in the industrial chain of industrial robots are, in order, robot unit product manufacturers, robot system integrators and industrial automation integrators. Most domestic robot enterprises are system integrators. Compared with the supplier of unit products, the system integrator should have product design capability and project experience, and provide standardized and personalized sets of equipment that can adapt to various application fields on the basis of a deep understanding of the user industry.

Vocabulary

policy /'pɒləsi/ n. 政策，方针；保险单

comprehensive /kɒmprɪ'hensɪv/ adj. 综合的；广泛的；有理解力的

implementation /ˌɪmplɪmen'teɪʃ(ə)n/ n. ［计］实现；履行；安装启用

cultivate /'kʌltɪveɪt/ vt. 培养；陶冶；耕作

emphasis /'emfəsɪs/ n. 重点；强调；加强语气

realistic /rɪə'lɪstɪk/ adj. 现实的；现实主义的；逼真的；实在论的

foundation /faʊn'deɪʃ(ə)n/ n. 基础；地基；基金会；根据；创立

aerospace /'eərəspeɪs/ n. 航空宇宙；［航］航空航天空间

ontology /ɒn'tɒlədʒɪ/ n. 本体论；存在论；实体论

breakthrough /'breɪkθruː/ n. 突破；突破性进展

integrator /'ɪntɪɡreɪtə/ n. ［自］积分器；［电子］积分电路；整合之人

biological /baɪə(ʊ)'lɒdʒɪk(ə)l/ adj. 生物的；生物学的

agricultural /ˌæɡrɪ'kʌltʃərəl/ adj. 农业的；农艺的

innovation /ˌɪnə'veɪʃn/ n. 创新，革新；新方法

Notes and analysis

Question 1：What are main areas mentioned in the policy?

Answer：_____

Question 2：What is the relationship between the robot and robot policy?

Answer：_____

Translation

机器人政策

综合考虑未来国际发展趋势和我国工业发展的现实基础条件，根据走中国特色工业化道路和加快转变经济发展方式的总体要求，我国需要加快实现由工业大国向工业强国的转变。我国的目标是，通过努力实现中国制造向中国创造、中国速度向中国质量、中国产品向中国品牌三大转变，基本实现工业化，进而成为制造业强国，如图 22-1 所示。

中国重点推进的十大领域建设分别是信息技术、数控工具和机器人、航空航天设备、海洋工程设备和高科技船舶、铁路设备、节能新能源汽车、电力设备、新材料、生物医药和医疗器械以及农业机械。

机器人的发展包含两个方向，一是开发工业机器人本体和关键零部件系列化产品，推动工业机器人产业化及应用，满足我国制造业转型升级的迫切需求。二是突破智能机器人关键技术，开发一批智能机器人，积极应对新一轮科技革命和产业变革的挑战。

中国工业机器人装机量为 13.5 万台（2018 年），约占全球的 1/3。目前，中国的机器人制造商正在扩大其在国内市场的份额。从产业链角度看，位于工业机器人产业链上的企业依次是机器人单元产品制造商、机器人系统集成商和工业自动化集成商。国内的机器人企业多为系统集成商。与单元产品的供应商相比，系统集成商还要具有产品设计能力和项目经验，并在对用户行业深刻理解的基础之上，提供可适应各种不同应用领域的标准化、个性化成套装备。

Appendix A   Resume

A resume (also named curriculum vitae, Fig. A-1) is presented in a formal and strategic format, which includes personal professional credentials. In other words, a resume is an important tool to market your candidature and present yourself on paper. A resume should include the career objective, qualifications, education background, working experience, technical skills, working history, etc. Once you have prepared your resume, the career objective is very important, as it makes the first impression on the employers and helps them to understand your expectations for the job.

Fig. A-1   Resume

A resume example:

Cosmin Hodgson

3216 No. 2 Highroad, Highton, New York, US, +63 23234567, cosmin. H@ yahoo. com

---

Objective

A career position in the maintenance of industrial robot.

Summary of Qualifications

Engineering work experience in automotive company.

Experience with the installation of industrial robot.

Experience with custom service.

Language certification.

Education

Bachelor of Mechanical Engineering, State University, New York, US

Overall GPA: 3. 2   May 2015

Industrial engineering experience

May 2015-June 2017, Industrial and Manufacturing engineering Co-op.

Designed and implemented a new assembly line with industrial robot.

Reorganized the inventory system.

Completed various industrial and manufacturing projects to support the assembly line with industrial robot.

（续）

July 2017-October 2019, Senior Project.

Developed a Support System by software tool.

Updated and maintained a new assembly line with industrial robot.

Developed and recommended improvements to an existing workstation.

Technical skill

Created a working database for given production systems.

Proficient using: Microsoft Excel, Word, PowerPoint, CAD/CAM systems.

Working History

Server-Woolworths Supermarket, 20 hours per week while attending school full-time.

Appendix B   Recruitment

From a standard recruitment, the employee candidate would catch the requirement information, such as the positions, roles, skill requirements, responsibilities, etc. The following example is a recruitment advertisement.

- Faculty Position

Positions available

The Chen's Industrial Automation welcomes qualified applicants to apply for the following open faculty positions:

Research and development engineer

- Company Information

Chen's Industrial Automation business offers a broad range of solutions for process and hybrid industries, including industry-specific integrated automation, electrification and digital solutions, control technologies, software and advanced services, as well as measurement & analytics, and marine and turbocharging offerings. Chen's Industrial Automation is No. 2 in the global market. Working closely with customers, Chen's Industrial Automation business is writing the future of safe and smart operations.

- Faculty Positions available

Area:

Industrial robot and data science

Qualifications:

Research and development engineer

Possess a PhD degree in engineering or related fields.

Have outstanding research accomplishments.

Have ability or show potential to lead the development of the related discipline.

Duties:

Research and development engineer

Salary and Benefits:

Internationally competitive salary with start-up grants and moving cost.

An environment conducive to active working.

Ample opportunities for career development.

Children education.

Applications:

Cover letter

Curriculum Vitae

Working statement

Research statement

Other support documents (for example, certificate of awards, etc.)

Contact:

Mr. Xiu Chen (H. R. Personnel of Chen's Industrial Automation)

Email address: xiu. c@ chen. com

Tel: +86-10-88668866

Appendix C   Application letter

How to reply a recruitment? The answer is a Job Application Letter (Fig. C-1). The Job Application Letter is your first interview. When applying for a job by mail, a job application letter must accompany your resume. Surveys of personnel directors of the 500 largest organizations showed that the vast majority (over 80%) have agreed or strongly agreed that they want to know:

Your personality. What do you like and what will you be like as an employee?

Why did you choose to apply for a job with this particular company?

What job are you specifically seeking?

What makes you feel that your education or experience relates to that job?

Fig. C-1   A job application letter

Here is an example of a job application letter.

12 Zhineng Avenue, Wuzhong, Soochow 215100

May 8, 2018

Ms. Jian Li

Director of Human and Resource

×××Corporation?

54 North Two Street, Wujiang, Soochow 215100

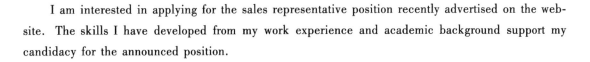

Dear Ms. Li,

I am interested in applying for the sales representative position recently advertised on the website. The skills I have developed from my work experience and academic background support my candidacy for the announced position.

As you can see from my resume, ABC Corporation provided an opportunity for me to gain practical experience with account maintenance and cold-calling new accounts. In addition, I have worked as a waiter for the past four years, learning firsthand how to effectively deal with customers and their needs. I have been formally praised by management several times, and was named "Employee of the Month."

I would very much hope to have an opportunity to discuss your specific needs and my overall capabilities for the announced position. You can reach me at (+86) -521-88665231. Thank you for considering me for this position.

Sincerely,

John K. Alberts

Appendix D　Mind Mapping

Mind mapping is a way to collect information inside and outside of your brain in a high efficient way. Basically, mind mapping is a note-making method which is creative and logical to "map out" your thoughts.

It can be found in Fig. D-1 that the Mind Map has a well-organized structure, including the topic center, main ideas and keywords. A lot of lines, symbols, text, colors and images are adopted to realize a colorful, memorable and highly organized diagram.

The great thing about mind mapping is that you can put your ideas down in any order, as soon as they pop into your mind. You are not constrained by thinking in order. Simply, throw out any or all ideas, then worry about reorganizing them later.

The mind map is the external mirror of your own radiation or natural thinking facilitated by a powerful graphic process, which provides the universal key to unlock the dynamic potential of the brain.

The essential characteristics of Mind Mapping are consisted of main idea, branches, keywords, and structure. When you start to make a Mind Mapping, the following steps should be considered. Firstly, it is necessary to determine the main theme and write it down in the center of the page, i. e. Industrial Robot (Fig. D-1). Secondly, the sub-themes are figured out with the branches drawing from the center to make a spider web, i. e. Introduction, Structure, and Component, etc. It is better to use single word or short phrase to make the illustration, and images are useful to stimu-

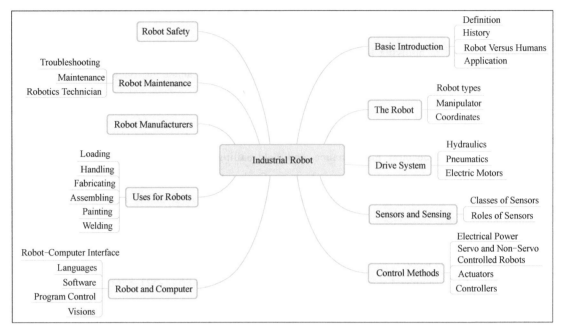

Fig. D-1   A mind map of industrial robot

late thought.

Appendix E   Three Laws of Robotics

What are the three laws of robotics? In 1950, Asimov published "I, robot", and added a subtitle 'Three Laws of Robotics' to the introduction. The three laws have a certain guiding significance for the creation of later generations.

Law Ⅰ: a robot may not injure a human being or, through inaction, allow a human being to come to harm.

Law Ⅱ: a robot must obey orders given to it by human beings except where such orders would conflict with the First Law.

Law Ⅲ: a robot must protect its own existence as long as such protection does not conflict with the First or Second Law.

All the robots are designed to follow the three guidelines, and breaking them will cause irreparable psychological damage. Nevertheless, damage is inevitable in some cases. When two people hurt each other, the robot cannot let one person to be hurt without any treatment, but it will cause harm to the other one, which causes the self-destruction of the robot.

The three laws shine in science fiction, as do the robots in other authors' science fiction. At the same time, the three laws also have a certain practical significance. Based on the three laws, a new discipline "mechanical ethics" is established to study the relationship between human beings and machines. Although the three laws have not been applied in the real robotics industry, many technical experts in the field of artificial intelligence and robotics also agree with these rules, and with the development of technology, the three laws may become the safety rules for robots in the future.

Appendix F   Multiples/sub-multiples abbreviations

Table F-1   Multiples/sub-multiples abbreviations

Abbreviation	Name	Relationship
Y	yotta	10^{24}
Z	zetta	10^{21}
E	exa	10^{18}
P	peta	10^{15}
T	tera	10^{12}
G	giga	10^{9}
M	mega or meg	10^{6}
k	kilo	10^{3}
h	hecto	10^{2}
da	deca	10
d	deci	10^{-1}
c	centi	10^{-2}
m	milli	10^{-3}
μ	micro	10^{-6}
n	nano	10^{-9}
p	pico	10^{-12}
f	femto	10^{-15}
a	atto	10^{-18}
z	zepto	10^{-21}
y	yocto	10^{-24}

Appendix G   Mathematic representation

Table G-1   Mathematic representation

Term	Symbol
+	Plus
−	Minus
×	Multiple
÷	Divide
=	Equal to
≈	Approximately equal to

（续）

Term	Symbol
∝	Proportional to
∞	Infinity
Σ	Sum of
Δ/d	Increment or finite difference operator
>	Greater than
<	Less than
e	Base of natural logarithms
$\log x$	Common logarithm of x
$\ln x$	Natural logarithm of x

Appendix H   Greek letters

Table H-1   Greek letters

Letter	Capital	Lower case
Alpha	A	α
Beta	B	β
Delta	Δ	γ
Epsilon	E	ε
Eta	H	η
Theta	Θ	θ
Lambda	Λ	λ
Mu	M	μ
Pi	Π	π
Rho	P	ρ
Sigma	Σ	σ
Phi	Φ	φ
Psi	Ψ	ψ
Omega	Ω	ω

Vocabulary

A

absorb /əb'zɔːb; -'sɔːb/ *vt.* 吸收；吸引；承受

absorber /əb'sɔːbə/ *n.* 减震器；吸收器；吸收体

academia /ˌækə'diːmɪə/ *n.* 学术界；学术生涯

accelerate /æk'seləreɪt/ *vt.* 使……加快；使……增速 *vi.* 加速；促进；增加

accomplish /ə'kʌmplɪʃ; ə'kɒm-/ *vt.* 完成；实现；达到

accumulation /əkjuːmjʊ'leɪʃ(ə)n/ *n.* 积聚，累积；堆积物

accuracy /'ækjʊrəsɪ/ *n.* [数] 精确度，准确性

accurate /'ækjərət/ *adj.* 精确的

achieve /ə'tʃiːv/ *vt.* 取得；成功 *vi.* 达到预期的目的

acquisition /ˌækwɪ'zɪʃ(ə)n/ *n.* 获得物，获得；收购

activity /æk'tɪvɪtɪ/ *n.* 活动；行动；活跃

actuator /'æktjʊeɪtə/ *n.* [自] 执行机构；激励者；促动器

adaptability /əˌdæptə'bɪlətɪ/ *n.* 适应性；可变性；适合性

additional /ə'dɪʃ(ə)n(ə)l/ *adj.* 附加的，额外的

adjustment /ə'dʒʌs(t)m(ə)nt/ *n.* 调整，调节；调节器

adopt /ə'dɒpt/ *vi.* 采取；过继 *vt.* 采取；接受；收养；正式通过

advantage /əd'vɑːntɪdʒ/ *n.* 优势；有利条件 *vt.* 有利于；使处于优势 *vi.* 获利

aerospace /'eərəspeɪs/ *n.* 航空宇宙；[航] 航空航天空间

aforementioned /əfɔː'menʃənd/ *adj.* 上述的；前面提及的

aggregate /'ægrɪgeɪt/ *n.* 合计；集合体 *adj.* 聚合的 *v.* 集合；聚集；合计

agricultural /ægrɪ'kʌltʃərəl/ *adj.* 农业的；农艺的

aim /eɪm/ *v.* 目的在于；引导；把……对准；瞄准 *n.* 目的；瞄准

alert /ə'lɜːt/ *n.* 警戒，警报 *v.* 使警觉，警告 *adj.* 警惕的，警觉的

algorithm /'ælgə'rɪð(ə)m/ *n.* [计] [数] 算法，运算法则

alternating current（AC） *n.* [电] 交流电

alternator /'ɔːltəneɪtə; 'ɒl-/ *n.* [电] 交流发电机

analog /'ænəlɒg/ *n.* [自] 模拟；类似物 *adj.* [自] 模拟的；有长短针的

anchor /'æŋkə/ *n.* 锚；抛锚停泊 *vt.* 抛锚；使固定；*vi.* 抛锚

angle /'æŋg(ə)l/ *v.* 斜移；谋取 *n.* 角，角度；视角；立场；角铁

animation /ænɪ'meɪʃ(ə)n/ *n.* 活泼；激励；动画，卡通片绘制

application /ˌæplɪ'keɪʃ(ə)n/ *n.* 应用；申请；应用程序

appropriate /ə'prəʊprɪət/ *adj.* 适当的；恰当的；合适的 *vt.* 占用，拨出

architecture   /'ɑːkɪtektʃə/ *n.* 建筑学；建筑风格；建筑式样；架构

arithmetic   /ə'rɪθmətɪk/ *n.* 算术，算法

array   /ə'reɪ/ *n.* 数组，阵列；列阵；一系列；衣服 *vt.* 排列，部署；打扮

arithmetic   /ə'rɪθmətɪk/ *n.* 算术，算法

arithmetic logic unit（ALU）   *n.* 算术逻辑单元

arm   /ɑːm/ *n.* 手臂；武器；袖子；装备；部门 *vt.* 武装；备战 *vi.* 武装起来

artificial   /ˌɑːtɪ'fɪʃl/ *adj.* 人造的；仿造的；非原产地的；武断的

aspect   /'æspekt/ *n.* 方面；方向；形势；外貌

assembly   /ə'semblɪ/ *n.* 装配；集会，集合 *n.* 汇编，编译

assign   /ə'saɪn/ *vt.* 分配；指派；［计］［数］赋值 *vi.* 将财产过户

assist   /ə'sɪst/ *v.* 参加，出席；有助益 *n.* 帮助；助攻

assurance   /ə'ʃʊər(ə)ns/ *n.* 保证，担保；（人寿）保险；确信；断言；

asymmetry   /eɪ'sɪmɪtrɪ/ *n.* 不对称

atmosphere   /'ætməsfɪə/ *n.* 气氛；大气；空气

attitude   /'ætɪtjuːd/ *n.* 态度；看法；意见；姿势

authentic   /ɔː'θentɪk/ *adj.* 真正的，真实的；可信的

automated   /'ɔːtəˌmeɪtɪd/ *adj.* 自动化的；机械化的 *v.* 自动操作

avoidance   /ə'vɔɪdəns/ *n.* 避免，逃避；废止

B

bankruptcy   /'bæŋkrʌptsɪ/ *n.* 破产

barrier   /'bærɪə/ *n.* 障碍物，屏障；界线 *vt.* 把…关入栅栏

base   /beɪs/ *n.* 基底；基础；基地 *v.* 以……作基础

battery   /'bætrɪ/ *n.* ［电］电池，蓄电池

belt   /belt/ *n.* 带；腰带；地带 *vt.* 用带子系住；用皮带抽打

beneficiary   /benɪ'fɪʃ(ə)rɪ/ *n.* ［金融］受益人，受惠者 *adj.* 拥有封地的

biological   /baɪə(ʊ)'lɒdʒɪk(ə)l/ *adj.* 生物的；生物学的

bond   /bɒnd/ *n.* 结合；约定；黏合剂 *vt.* 使结合；以…作保 *vi.* 结合

brain   /breɪn/ *n.* 头脑，智力；脑袋 *vt.* 猛击…的头部

brake   /breɪk/ *n.* 刹车；阻碍 *v.* 刹车；阻碍

brand   /brænd/ *v.* 打烙印；加商标于 *n.* 品牌，商标；类型；烙印

breakthrough   /'breɪkθruː/ *n.* 突破；突破性进展

brightness   /'braɪtnɪs/ *n.* ［光］［天］亮度；聪明，活泼；鲜艳；愉快

buffer   /'bʌfə/ *n.* ［计］缓冲区；缓冲器，［车辆］减震器 *vt.* 缓冲

C

cabinet   /'kæbɪnɪt/ *n.* 内阁；橱柜 *adj.* 内阁的；私下的

camera   /'kæm(ə)rə/ *n.* 照相机；摄影机

capability   /ˌkeɪpə'bɪlətɪ/ *n.* 才能，能力；性能，容量

Industrial Robot

capacity /kə'pæsɪtɪ/ *n.* 能力；容量；资格，地位；生产力

Cartesian coordinate *n.* 笛卡儿坐标

cast /kɑ:st/ *n.* 投掷，抛；*vt.* 投，抛；浇铸 *vi.* 投，计算

catalog /'kætəlɒg/ *n.* ［图情］［计］目录；登记 *vt.* 登记；为…编目录 *vi.* 编目录

category /'kætɪg(ə)rɪ/ *n.* 种类，分类；［数］范畴

ceiling /'si:lɪŋ/ *n.* 天花板；上限

centerline /'sentəlain/ *n.* 中心线

central processing unit *n.* ［计］中央处理机；中央处理单元

chain /tʃeɪn/ *n.* 链；束缚；枷锁 *vt.* 束缚

characteristic /kærəktə'rɪstɪk/ *n.* 特征；特性；特色 *adj.* 典型的；表示特性的

chart /tʃɑ:t/ *n.* 图表；图纸；排行榜 *vt.* 绘制…的图表；记录

chip /tʃɪp/ *n.* 芯片，晶片；碎片 *v.* 打缺，弄缺；铲，凿，削

circuit /'sɜ:kɪt/ *n.* ［电子］电路，回路 *vt.* 绕回…环行 *vi.* 环行

circuitry /'sɜ:kɪtrɪ/ *n.* 电路；电路系统；电路学；一环路

circumstance /'sɜ:kəmstəns/ *n.* 环境；状况；境遇；命运 *vt.* 处于某种情况

classify /'klæsɪfaɪ/ *vt.* 分类；分等

CNC *n.* 电脑数值控制（Computer Numerical Control）

coating /'kəʊtɪŋ/ *n.* 涂层；包衣；衣料 *v.* 给…穿上外衣

code /kəʊd/ *n.* 代码，密码；编码；法典 *vt.* 编码 *vi.* 指定遗传密码

collaborative /kə'læbərətɪv/ *adj.* 合作的，协作的

collation /kə'leɪʃ(ə)n/ *n.* 校对

color /'kʌlə(r)/ *n.* 颜色；肤色；颜料 *vt.* 粉饰；歪曲 *vi.* 变色；获得颜色

column /'kɒləm/ *n.* 纵队，列；专栏；圆柱，柱形物

combination /kɒmbɪ'neɪʃ(ə)n/ *n.* 结合；组合；联合；［化学］化合

combustion /kəm'bʌstʃ(ə)n/ *n.* 燃烧，氧化；骚动

commination /ˌkɒmɪ'neɪʃ(ə)n/ *n.* 威吓

common /'kɒmən/ *n.* 普通；平民；公有地 *adj.* 共同的；普通的

communication /kəmju:nɪ'keɪʃ(ə)n/ *n.* 通讯，［通信］通信；交流；信函

community /kə'mju:nətɪ/ *n.* 社区；［生态］群落；共同体；团体

compact /kəm'pækt/ *n.* 契约 *adj.* 紧凑的；坚实的 *v.* 把……压实；使简洁

company /'kʌmp(ə)nɪ/ *n.* 公司；陪伴，同伴；连队 *vt.* 陪伴 *vi.* 交往

competitiveness /kəm'petətɪvnɪs/ *n.* 竞争力

component /kəm'pəʊnənt/ *adj.* 组成的；构成的 *n.* 组成部分；成分；元件

comprehensive /kɒmprɪ'hensɪv/ *n.* 综合学校 *adj.* 综合的；广泛的

compressed /kəm'prest/ *adj.* （被）压缩的；扁的 *v.* （被）压紧，精简

compressor /kəm'presə/ *n.* 压缩机；压缩物；收缩肌；［医］压迫器

concept /'kɒnsept/ *n.* 观念，概念

consists of 包含；由…组成；充斥着

constant /'kɒnst(ə)nt/ *n.* ［数］常数；恒量 *adj.* 不变的；恒定的；经常的

construction　/kənˈstrʌkʃ(ə)n/ *n.* 建设；建筑物；解释；造句

consumer　/kənˈsjuːmə/ *n.* 消费者；用户，顾客

contact　/ˈkɒntækt/ *n.* 接触，联系 *vt.* 使接触，联系 *vi.* 使接触，联系

containment　/kənˈteɪnm(ə)nt/ *n.* 控制，抑制；遏制；封锁（政策）

content　/kənˈtent/ *n.* 内容，目录；容量 *adj.* 满意的 *vt.* 使满足

controller　/kənˈtrəʊlə/ *n.* 控制器；管理员

convert　/kənˈvɜːt/ *vt.* 使转变；转换…；使…改变信仰 *vi.* 转变，变换

conveyor　/kənˈveɪə/ *n.* 输送机，[机] 传送机；传送带；运送者，传播者

coordinate transformation　*n.* [数] 坐标变换

coordination　/kəʊˌɔːdɪˈneɪʃən/ *n.* 协调，调和；对等，同等

core　/kɔː/ *n.* 核心；要点；果心；[计] 磁心 *vt.* 挖…的核

corrosion　/kəˈrəʊʒən/ *n.* 腐蚀；腐蚀产生的物质；衰败

count　/kaʊnt/ *v.* 数数；计算总数 *n.* 总数；数数；量的计数

counter　/ˈkaʊntə/ *n.* 柜台；计数器 *vi.* 逆向移动，对着干；反驳 *adj.* 相反的

counterclockwise　/kaʊntəˈklɒkwaɪz/ *adj.* 反时针方向的 *adv.* 反时针方向

creator　/kriːˈeɪtə/ *n.* 创造者；创建者

critical　/ˈkrɪtɪk(ə)l/ *adj.* 鉴定的；[核] 临界的；批评的，决定性的；评论的

cultivate　/ˈkʌltɪveɪt/ *vt.* 培养；陶冶；耕作

customize　/ˈkʌstəmaɪz/ *vt.* 定做，按客户具体要求制造

cut　/kʌt/ *v.* 割破；切下；剪切 *n.* 切，割 *adj.* 缩减的；割下的

cycle　/ˈsaɪk(ə)l/ *n.* 循环；周期 *vt.* 使循环；使轮转 *vi.* 循环；轮转

cylindrical coordinate　*n.* [数] 柱面坐标，圆柱坐标

D

data　/ˈdeɪtə/ *n.* 数据；资料

debris　/ˈdeɪbriː/ *n.* 碎片，残骸

debug　/diːˈbʌg/ *vt.* 调试；除错，改正有毛病部分

deburr　/diːˈbə/ *v.* 抛光，修边；去除毛边；除去脏物

decelerate　/diːˈseləreɪt/ *vt.* 使减速 *vi.* 减速，降低速度

dedicate　/ˈdedɪkeɪt/ *vt.* 致力；献身；题献

deep　/diːp/ *adj.* 深的 *adv.* 在深处，深深地；边线地

deformation　/ˌdiːfɔːˈmeɪʃ(ə)n/ *n.* 变形

demonstrate　/ˈdemənstreɪt/ *vt.* 证明；展示；论证 *vi.* 示威

demonstration　/ˌdemənˈstreɪʃ(ə)n/ *n.* 示范；证明；示威游行

department　/dɪˈpɑːtm(ə)nt/ *n.* 部；部门；系；科；局

deposit　/dɪˈpɒzɪt/ *n.* 存款；订金；沉淀物 *vt.* 使沉积；存放 *vi.* 沉淀

description　/dɪˈskrɪpʃ(ə)n/ *n.* 描述，描写；类型；说明书

design　/dɪˈzaɪn/ *v.* 设计，构思；计划 *n.* 设计；构思；设计图样

detail　/ˈdiːteɪl/ *n.* 细节，琐事；具体信息 *vt.* 详述；选派 *vi.* 画详图

determine /dɪ'tɜːmɪn/ v. (使) 下决心 vt. 决定，确定；

device /dɪ'vaɪs/ n. 装置；策略；图案；设备；终端

diagnose /'daɪəgnəʊz;-'nəʊz/ vt. 诊断；断定 vi. 诊断；判断

die-casting n. 压模法；铸造法

digital /'dɪdʒɪt(ə)l/ n. 数字；键 adj. 数字的；手指的

direct current (DC) n. [电] 直流电

disadvantage /dɪsəd'vɑːntɪdʒ/ n. 缺点；不利条件；损失

disassembly /ˌdɪsə'semblɪ/ n. 拆卸；分解

discrimination /dɪˌskrɪmɪ'neɪʃ(ə)n/ n. 歧视；区别，辨别；识别力

distance /'dɪstəns/ n. 距离；远方；疏远；间隔 vt. 疏远；把…远远甩在后面

distribute /dɪ'strɪbjuːt/ vt. 分配；散布；分开；把…分类

distribution /dɪstrɪ'bjuːʃ(ə)n/ n. 分布；分配；供应

division /dɪ'vɪʒ(ə)n/ n. [数] 除法；部门；分配；师（军队）；赛区

doubt /daʊt/ n. 怀疑；疑问；疑惑 v. 怀疑；不信

downtime /'daʊntaɪm/ n. 停工期；[电子] 故障停机时间

dozen /'dʌz(ə)n/ n. 十二个，一打 adj. 一打的

draw /drɔː/ n. 平局；抽签 vi. 拉；拖 vt. 画；拉；吸引

drill /drɪl/ n. 钻子 vt. 钻孔；训练；条播 vi. 钻孔；训练

drive /draɪv/ v. 开车；推动；驱赶；迫使，逼迫

dust /dʌst/ n. 灰尘；尘埃；尘土 vt. 撒；拂去灰尘 vi. 拂去灰尘；化为粉末

duty /'djuːtɪ/ n. 责任；[税收] 关税；职务

dynamic /daɪ'næmɪk/ n. 动态；动力 adj. 动态的；动力的；动力学的

E

education /ˌedʒʊ'keɪʃn/ n. 教育；培养；教育学

efficiency /ɪ'fɪʃ(ə)nsɪ/ n. 效率；效能；功效

elbow /'elbəʊ/ n. 肘部；弯头；扶手 vt. 推挤；用手肘推开

electric /ɪ'lektrɪk/ n. 电；电气车辆；带电体 adj. 电的；电动的；发电的

electrification /ɪˌlektrɪfɪ'keɪʃən/ n. 电气化；带电；充电

electronic /ɪˌlek'trɒnɪk/ adj. 电子的 n. 电子电路；电子器件

eliminate /ɪ'lɪmɪneɪt/ vt. 消除；排除

elimination /ɪˌlɪmɪ'neɪʃən/ n. 消除；淘汰；除去

emergency /ɪ'mɜːdʒ(ə)nsɪ/ n. 紧急情况；突发事件 adj. 紧急的；备用的

emphasis /'emfəsɪs/ n. 重点；强调；加强语气

enable /ɪn'eɪb(ə)l;en-/ v. 使能够；使成为可能；（计算机）启动

encounter /ɪn'kaʊntə;en-/ v. 遭遇；邂逅；遇到 n. 遭遇；偶然碰见

energy /'enədʒɪ/ n. [物] 能量；精力；活力；精神

engine /'endʒɪn/ n. 引擎，发动机；机车，工具

engineer /ˌendʒɪ'nɪə/ n. 工程师 vt. 设计；策划；精明地处理 vi. 设计；建造

ensure   /ɪnˈʃɔː;-ˈʃʊə;en-/ *vt.* 保证，确保；使安全

enterprise   /ˈentəpraɪz/ *n.* 企业；事业；进取心；事业心

entire   /ɪnˈtaɪə;en-/ *adj.* 全部的，整个的；全体的

environment   /ɪnˈvaɪrənmənt/ *n.* 环境，外界

era   /ˈɪərə/ *n.* 时代；年代；纪元

error   /ˈerə/ *n.* 误差；错误；过失

essence   /ˈes(ə)ns/ *n.* 本质，实质；精华

essential   /ɪˈsenʃ(ə)l/ *n.* 本质；要素；必需品 *adj.* 基本的；必要的

establish   /ɪˈstæblɪʃ;e-/ *v.* 建立，创立；确立；获得接受；查实，证实

exaggerate   /ɪgˈzædʒəreɪt;eg-/ *vt.* 使扩大；使增大 *vi.* 夸大；夸张

examine   /ɪgˈzæmɪn;eg-/ *vt.* 检查；调查；检测；考试 *vi.* 检查；调查

exchange   /ɪksˈtʃeɪndʒ/ *n.* 交换；交流；兑换 *vt./vi.* 交换；交易；兑换

execute   /ˈeksɪkjuːt/ *vt.* 实行；执行

exhaust   /ɪgˈzɔːst;eg-/ *n.* 废气；排气管 *v.* 使筋疲力尽；耗尽

existing   /ɪgˈzɪstɪŋ/ *v.* 存在；被发现 *adj.* 存在的；现行的

explorer   /ekˈsplɔːrə(r)/ *n.* 探险家；勘探者；探测器；[医] 探针

exposition   /ekspəˈzɪʃ(ə)n/ *n.* 博览会；阐述；展览会

exposure   /ɪkˈspəʊʒə;ek-/ *n.* 暴露；曝光；揭露；陈列

expression   /ɪkˈspreʃən/ *n.* 表现，表示，表达；表情，态度

extensibility   /ikˌstensəˈbiləti/ *n.* 展开性；可延长性

extension   /ɪkˈstenʃ(ə)n;ek-/ *n.* 延长；延期；扩大；伸展；电话分机

external   /ɪkˈstɜːn(ə)l;ek-/ *n.* 外部；外观 *adj.* 外部的；表面的；外国的

externally   /eksˈtəːnəli/ *adv.* 外部地；外表上，外形上

extra   /ˈekstrə/ *n.* 额外的事物 *adv.* 额外；特别地 *adj.* 额外的；特大的

extract   /ˈekstrækt;ɪkˈstrækt/ *n.* 摘录；汁 *v.* 提炼；选取，摘录；取出

extreme   /ɪkˈstriːm;ek-/ *n.* 极端；最大程度 *adj.* 极端的；极度的；偏激的

F

fabrication   /fæbrɪˈkeɪʃ(ə)n/ *n.* 制造，建造；装配；伪造物

failure   /ˈfeɪljə/ *n.* 失败；故障；失败者；破产

false   /fɔːls;fɒls/ *adj.* 错误的；虚伪的；伪造的 *adv.* 欺诈地

fatigue   /fəˈtiːg/ *n.* 疲劳，疲乏；杂役 *adj.* 疲劳的 *vt.* 使疲劳 *vi.* 疲劳

fault   /fɔːlt/ *n.* 故障；[地质] 断层；错误 *vi.* 弄错；产生断层

feature   /ˈfiːtʃə(r)/ *n.* 特色，特征 *vt.* 特写；以…为特色 *vi.* 起重要作用

feedback   /ˈfiːdbæk/ *n.* 反馈；资料；回复

field work   *n.* 现场工作

filter   /ˈfɪltə/ *v.* 过滤；渗透；用过滤法除去；缓行 *n.* 过滤器；筛选程序

flat belts   *n.* 平型传动带

flexible   /ˈfleksəb(ə)l/ *adj.* 灵活的；柔韧的；易弯曲的

Industrial Robot

fluctuation /ˌflʌktʃʊˈeɪʃ(ə)n;-tjʊ-/ *n.* 起伏，波动

fluid /ˈfluːɪd/ *adj.* 流动的；不固定的，易变的；液压传动的 *n.* 流体，液体

focus /ˈfəʊkəs/ *n.* 焦点；中心；焦距 *vt.* 使集中 *vi.* 集中；调节焦距

force /fɔːs/ *n.* 力量；武力；军队 *vt.* 促使，推动；强迫；强加

foresee /fɔːˈsiː/ *vt.* 预见；预知

format /ˈfɔːmæt/ *n.* 格式；版式；开本 *vt.* 使格式化 *vi.* 设计版式

format /ˈfɔːmæt/ *n.* 格式；版式 *vt.* 使格式化；规定…的格式 *vi.* 设计版式

formula /ˈfɔːmjʊlə/ *n.* ［数］公式，准则；配方；婴儿食品

foundation /faʊnˈdeɪʃ(ə)n/ *n.* 基础；地基；基金会；根据；创立

fracture /ˈfræktʃə/ *n.* 破裂，断裂 *vt.* 使破裂 *vi.* 破裂；折断

frame /freɪm/ *n.* 框架；结构 *adj.* 有构架的 *vt.* 给（图画或照片）配框；设计

frictional /ˈfrɪkʃənl/ *adj.* ［力］摩擦的；由摩擦而生的

functionality /fʌŋkʃəˈnælətɪ/ *n.* 功能；［数］泛函性，函数性

fundamental /fʌndəˈment(ə)l/ *n.* 基本原理；基本原则 *adj.* 基本的，根本的

furniture /ˈfɜːnɪtʃə/ *n.* 家具；设备；储藏物

G

gap /gæp/ *n.* 间隙；缺口；差距；分歧 *vt.* 使形成缺口 *vi.* 裂开

gather /ˈgæðə/ *n.* 聚集 *vt.* 收集；使…聚集 *vi.* 聚集；皱起

gauge /geɪdʒ/ *n.* 计量器；标准尺寸；容量规格 *vt.* 测量；估计；给…定规格

gear /gɪə/ *n.* 齿轮；传动装置 *vt.* 开动；搭上齿轮 *vi.* 适合；搭上齿轮

gear teeth *n.* 齿轮齿

genera /ˈdʒenərə/ *n.* ［生物］属（genus 的复数形式）；种；类

generation /dʒenəˈreɪʃ(ə)n/ *n.* 一代；产生；一代人；生殖

geometric /dʒɪəˈmetrɪk/ *adj.* 几何学的；［数］几何学图形的

Germany /ˈdʒɜːmənɪ/ *n.* 德国

global /ˈgləʊb(ə)l/ *adj.* 全球的；总体的；球形的

govern /ˈgʌv(ə)n/ *vt.* 管理；支配；统治；控制 *vi.* 进行统治

graph /grɑːf/ *n.* 图表；曲线图 *vt.* 用曲线图表示

graphic /ˈgræfɪk/ *adj.* 形象的；图表的；绘画似的

grasp /grɑːsp/ *v.* 抓牢，握紧 *n.* 抓，握；理解，把握；权力，控制

grid /grɪd/ *n.* 网格；格子，栅格；输电网

grind /graɪnd/ *n.* 磨；苦工作 *vt.* 磨碎；磨快 *vi.* 磨碎；折磨

gripper /ˈgrɪpə/ *n.* 夹子，钳子；抓器，抓爪

guardian /ˈgɑːdɪən/ *n.* ［法］监护人，保护人；守护者 *adj.* 守护的

guide /gaɪd/ *n.* 指南；向导；入门书 *vt.* 引导；带领；操纵 *vi.* 担任向导

H

halfway /ˈhɑːfweɪ/ *adj.* 中途的；不彻底的 *adv.* 到一半；在中途

hand /hænd/ n. 手，手艺；帮助；指针；插手 vt. 传递，交给；支持；搀扶

hardware /'hɑːdweə/ n. 计算机硬件；五金器具

harsh /hɑːʃ/ adj. 严厉的；严酷的

hearing /'hɪərɪŋ/ n. 听力；审讯，听讯 v. 听见

horizontal /hɒrɪ'zɒnt(ə)l/ n. 水平线，水平面 adj. 水平的；地平线的

humidity /hjuː'mɪdɪtɪ/ n. [气象] 湿度；湿气

hydraulic /haɪ'drɔːlɪk；haɪ'drɒlɪk/ adj. 液压的；水力的；水力学的

ideal /aɪ'dɪəl；aɪ'diːəl/ n. 理想；典范 adj. 理想的；完美的；想象的

illustration /ɪlə'streɪʃ(ə)n/ n. 说明；插图；例证；图解

imitate /'ɪmɪteɪt/ vt. 模仿，仿效；仿造，仿制

imitation /ɪmɪ'teɪʃ(ə)n/ n. 模仿，仿造；仿制品 adj. 人造的，仿制的

impact /'ɪmpækt/ vi. 影响；冲突 n. 影响；碰撞 vt. 挤入，压紧

implement /'ɪmplɪm(ə)nt/ n. 工具；手段 vt. 实施，执行；实现

implementation /ɪmplɪmen'teɪʃ(ə)n/ n. [计] 实现；履行；安装启用

improve /ɪm'pruːv/ vt. 改善，增进；提高…的价值 vi. 增加；变得更好

inaccessible /ɪnək'sesɪb(ə)l/ adj. 难达到的；难接近的；难见到的

index /'ɪndeks/ n. 指标；指数；索引；指针 vt. 指出；编入索引中 vi. 做索引

individual /ɪndə'vɪdʒʊəl/ adj. 单独的，独特的 n. 个人，个体

inelastic /ɪnɪ'læstɪk/ adj. 无弹性的；无适应性的；不能适应的

influence /'ɪnflʊəns/ n. 影响；势力；感化 vt. 影响；改变

information /ɪnfə'meɪʃ(ə)n/ n. 信息，资料；知识；情报

infrared /ɪnfrə'red/ adj. 红外线的；（设备、技术）使用红外线的

infrastructure /'ɪnfrəstrʌktʃə/ n. 基础设施；公共建设；下部构造

initial /ɪ'nɪʃ(ə)l/ n. 词首大写字母 adj. 最初的；字首的

initiative /ɪ'nɪʃɪətɪv；-ʃə-/ n. 主动权；新方案 adj. 主动的；自发的；起始的

injection /ɪn'dʒekʃ(ə)n/ n. 注射；注射剂；充血；射入轨道

innovation /ɪnə'veɪʃn/ n. 创新，革新；新方法

input/output（I/O） n. 输入/输出

inspection /ɪn'spekʃn/ n. 视察，检查

iinstallation /ɪnstə'leɪʃ(ə)n/ n. 安装，装置；就职

institute /'ɪnstɪtjuːt/ v. 实行，建立 n. 机构，研究所，学会

instruction /ɪn'strʌkʃ(ə)n/ n. 指令，命令；指示；教导；用法说明

instructor /ɪn'strʌktə/ n. 指导书；教员；指导者

integrate /'ɪntɪgreɪt/ n. 一体化 adj. 整合的 vi. 求积分 vt. 使…完整

integration /ɪntɪ'greɪʃ(ə)n/ n. 集成；综合

integrator /'ɪntɪgreɪtə/ n. [自] 积分器；[电子] 积分电路；整合之人

integrity /ɪn'tegrɪtɪ/ n. 完整；正直；诚实；廉正

Industrial Robot

intelligent /ɪnˈtelɪdʒ(ə)nt/ *adj.* 智能的；聪明的；理解力强的

interact /ˌɪntərˈækt/ *n.* 幕间剧；幕间休息 *vt.* 互相影响；互相作用

interactivity /ˈɪntəˈæktɪv/ *n.* 交互性；互动性

interconnection /ˌɪntəkəˈnekʃən/ *n.* ［计］互连；互相联络

interface /ˈɪntəfeɪs/ *n.* 界面；<计>接口；交界面 *v.* 接合，连接；［计算机］使联系

interior /ɪnˈtɪəriɪ(r)/ *n.* 内部，里面；本质 *adj.* 内部的，里面的

internal /ɪnˈtɜːn(ə)l/ *adj.* 内部的；本身的；内心的 *n.* 内脏；内部特征

interpolation /ɪnˌtɜːpəˈleʃən/ *n.* 插入；篡改；填写；插值

interpretation /ɪntɜːprɪˈteɪʃ(ə)n/ *n.* 解释；翻译；演出

intersection /ɪntəˈsekʃ(ə)n/ *n.* 交叉；十字路口；交集；交叉点

interval /ˈɪntəv(ə)l/ *n.* 间隔；间距；幕间休息

invariably /ɪnˈveəriəblɪ/ *adv.* 总是；不变地；一定地

involve /ɪnˈvɒlv/ *vt.* 包含；牵涉；使陷于；潜心于

issue /ˈɪʃuː；ˈɪsjuː/ *n.* 问题；发行物 *vt.* 发行，发布 *vi.* 发行；流出

Italy /ˈɪtəli/ *n.* 意大利

J

join /dʒɔɪn/ *n.* 结合；连接；接合点 *vi.* 加入；参加；结合 *vt.* 参加；结合

joint /dʒɔɪnt/ *adj.* 联合的，连接的 *n.* 关节；接合点 *v.* 连接，贴合

judgment /ˈdʒʌdʒmənt/ *n.* 判断；裁判；判决书；辨别力

K

keyboard /ˈkiːbɔːd/ *n.* 键盘 *vt.* 键入 *vi.* 用键盘进行操作

L

language /ˈlæŋgwɪdʒ/ *n.* 语言；语言文字；表达能力

laser /ˈleɪzə/ *n.* 激光

latter /ˈlætə/ *adj.* 后者的；近来的；后面的；较后的

launch /lɔːntʃ/ *v.* 发射，发动 *n.* 发射；发行

lens /lenz/ *n.* 透镜，镜头；眼睛中的水晶体；晶状体 *vt.* 给……摄影

linear /ˈlɪnɪə/ *adj.* 线的，线型的；直线的，线状的；长度的

link /lɪŋk/ *n.* ［计］链环，环节 *vt.* 连接；联合 *vi.* 连接起来

load /ləud/ *n.* 负载，负荷；装载量 *vt.* 使担负；装填 *vi.* ［力］加载；装载

loading /ˈləudɪŋ/ *n.* 装载；装货；装载的货 *v.* 装载，装填，装入

local /ˈləuk(ə)l/ *adj.* 当地的，局部的；局域的 *n.* 当地人；局部

logic /ˈlɒdʒɪk/ *n.* 逻辑；逻辑学；逻辑性 *adj.* 逻辑的

logistics /ləˈdʒɪstɪks/ *n.* ［军］后勤；后勤学

loop /luːp/ *v.* 使成环；环行 *n.* 环状物、圈；环状结构

loose /luːs/ *n.* 放纵 *adj.* 宽松的；不牢固的 *vt.* 释放 *adv.* 松散地

lubrication /ˌluːbrɪˈkeɪʃən/ n. 润滑；润滑作用

M

magnetic /mæɡˈnetɪk/ adj. 地磁的；有磁性的；有吸引力的

maintenance /ˈmeɪntənəns/ n. 维护，维修；保持；生活费用

manifestation /ˌmænɪfeˈsteɪʃ(ə)n/ n. 表现；显示；示威运动

manipulator /məˈnɪpjuleɪtə(r)/ n. 操纵器；操作者

manual /ˈmænju(ə)l/ n. 说明书；小册子 adj. 体力的；手控的

manufacture /ˌmænjuˈfæktʃə/ n. 制造；产品；制造业 vt. 制造；加工；vi. 制造

mate /meɪt/ n. 助手 vt. 使配对；使一致；结伴 vi. 成配偶；紧密配合

material /məˈtɪərɪəl/ adj. 物质的；客观存在的 n. 材料；用具

mathematics /mæθ(ə)ˈmætɪks/ n. 数学；数学运算

maximum /ˈmæksɪməm/ n. [数] 极大，最大限度 adj. 最高的；最大极限的

measurement /ˈmeʒəm(ə)nt/ n. 测量；[计量] 度量；尺寸；量度制

mechanical /mɪˈkænɪk(ə)l/ adj. 机械的；力学的；手工操作的

memory /ˈmem(ə)rɪ/ n. 记忆，记忆力；内存，[计] 存储器；回忆

merge /mɜːdʒ/ vt. 合并；使合并；吞没 vi. 合并；融合

mesh /meʃ/ n. 网眼；网丝；网格 vt. [机] 啮合 vi. 相啮合

metallic /mɪˈtælɪk/ adj. 金属的，含金属的

microcomputer /ˈmaɪkrə(ʊ)kɒmˌpjuːtə/ n. 微电脑；[计] 微型计算机

microelectronic /ˌmaɪkrəʊɪlekˈtrɒnɪk/ adj. [电子] 微电子的

microprocessor /maɪkrə(ʊ)ˈprəʊsesə/ n. [计] 微处理器

mill /mɪl/ v. 碾磨，切割（金属），铣 n. 磨坊，磨粉厂；机器，铣床

mobile /ˈməʊbaɪl/ n. 移动电话 adj. 可移动的；机动的；易变的；非固定的

mode /məʊd/ n. 模式；方式；风格；时尚

modern /ˈmɒd(ə)n/ n. 现代人；有思想的人 adj. 现代的，近代的；时髦的

mold /məʊld/ v. 浇铸，塑造 n. 模具；铸模；框架

molten /ˈməʊlt(ə)n/ adj. 熔化的；铸造的；炽热的 v. 换毛；脱毛

monitor /ˈmɒnɪtə/ n. 监视器；监听器；监控器；显示屏 vt. 监控

moral /ˈmɒr(ə)l/ n. 道德；寓意 adj. 道德的；精神上的；品行端正的

motion /ˈməʊʃ(ə)n/ n. 动作；移动；手势 vt. 运动 vi. 运动；打手势

movement /ˈmuːvm(ə)nt/ n. 运动；活动；运转；乐章

multifunctional /ˌmʌltiˈfʌŋʃənel/ adj. 多功能的

mutually /ˈmjuːtʃuəli/ adv. 互相地；互助

N

nanoscale n. 纳米级

noise /nɔɪz/ n. [环境] 噪音；响声；杂音 vt. 谣传 vi. 发出声音

noncontact /ˌnɒnˈkɒntækt/ n. 无触头，无触点 adj. 没有接触的

Industrial Robot

numerous   /ˈnjuːm(ə)rəs/ *adj.* 许多的，很多的

nursing   /ˈnɜːsɪŋ/ *n.* 护理；看护；养育 *v.* 看护；养育（nurse 的 ing 形式）

O

object   /ˈɒbdʒɪkt; -dʒekt/ *n.* 目标；物体；客体 *vi.* 反对；拒绝

obscure   /əbˈskjʊə/ *adj.* 不清楚的；隐蔽的 *vt.* 掩盖；隐藏

observer   /əbˈzɜːvə/ *n.* 观察者；[天] 观测者；遵守者

obstacle   /ˈɒbstək(ə)l/ *n.* 障碍，干扰，妨碍；障碍物

offline   /ɒfˈlaɪn/ *adj.* （计算机）未联网的；离线的 *adv.* 未连线地；脱机地

oil   /ɔɪl/ *n.* 油；石油 *vt.* 加油；使融化 *vi.* 融化；加燃油

ontology   /ɒnˈtɒlədʒɪ/ *n.* 本体论；存在论；实体论

operation   /ɒpəˈreɪʃ(ə)n/ *n.* 操作；经营；[外科] 手术；[数][计] 运算

operator   /ˈɒpəreɪtə/ *n.* 经营者；操作员；运营商

opportunity   /ɒpəˈtjuːnɪtɪ/ *n.* 时机，机会

organization   /ˌɔːɡənaɪˈzeɪʃn/ *n.* 组织；机构；体制；团体

oriented   /ˈɔːrɪentɪd/ *v.* 使朝向，使面对；确定方位 *adj.* 以……为方向的

origin   /ˈɒrɪdʒɪn/ *n.* 起源；原点；出身；开端

original   /əˈrɪdʒɪn(ə)l/ *adj.* 原来的；开始的；创新的；原作的 *n.* 原件

oscillator   /ˈɒsɪleɪtə(r)/ *n.* [电子] 振荡器

overload   /əʊvəˈləʊd/ *n.* 超载量 *v.* （使）过载，超载

P

package   /ˈpækɪdʒ/ *n.* 包，包裹；[计] 程序包 *vt.* 打包；将…包装

painting   /ˈpeɪntɪŋ/ *n.* 绘画；油画；着色 *v.* 绘画；涂色于

palletizing   /ˈpælə,taizɪŋ/ *n.* 夹板装载；码垛堆积 *v.* 把…装在货盘上

parallel   /ˈpærəlel/ *n.* 平行线；对比 *adj.* 平行的；类似的 *vt.* 使…与…平行

parameter   /pəˈræmɪtə/ *n.* 参数；系数；参量

pass   /pɑːs/ *v.* 通过，经过；传递 *n.* 及格；经过；通行证

patient   /ˈpeɪʃ(ə)nt/ *n.* 病人，患者；受动者，承受者 *adj.* 有耐心的，能容忍的

pattern   /ˈpæt(ə)n/ *n.* 模式；图案；样品 *vt.* 模仿；以图案装饰 *vi.* 形成图案

payload   /ˈpeɪləʊd/ *n.* 有效载荷，有效负荷；收费载重，酬载

perception   /pəˈsepʃ(ə)n/ *n.* 认识能力；知觉，感觉；洞察力

period   /ˈpɪərɪəd/ *n.* 周期，期间；时期；句号 *adj.* 某一时代的

peripheral   /pəˈrɪf(ə)r(ə)l/ *adj.* 外围的；次要的 *n.* 外部设备

personnel   /pɜːsəˈnel/ *n.* 人事部门；全体人员 *adj.* 人员的；有关人事的

perspective   /pəˈspektɪv/ *n.* 观点；远景；透视图 *adj.* 透视的

phenomena   /fəˈnɒmɪnə/ *n.* 现象

phenomenon   /fɪˈnɒmɪnən/ *n.* 现象；奇迹

photocell   /ˈfəʊtəʊsel/ *n.* [电] 光电池；[电子] 光电管

photoelectric /ˌfəutəuɪ'lektrɪk/ adj. ［电子］光电的

physics /'fɪzɪks/ n. 物理学；物理现象

picture /'pɪktʃə/ n. 照片；影片；景色 vt. 画；想象；描写

pipe /paɪp/ n. 管 vt. 用管道输送；用管乐器演奏 vi. 吹笛；尖叫

pit /pɪt/ n. 矿井；深坑 vt. 使竞争；使凹下 vi. 凹陷；起凹点

pixel /'pɪks(ə)l;-sel/ n. （显示器或电视机图像的）像素

pixel depth n. 像素深度

plane /pleɪn/ n. 飞机；平面；水平 adj. 平的 vt. 刨平 vi. 刨

plant /plɑ:nt/ n. 工厂，车间；植物 vt. 种植；培养；安置 vi. 种植

plastic /'plæstɪk/ adj. 塑料制的；人造的，塑性的 n. 塑料；塑料学

pneumatic /nju:'mætɪk/ n. 气胎 adj. 气动的；充气的；有气胎的

polar /'pəulə/ n. 极面；极线 adj. 极地的；两极的；正好相反的

polar coordinates n. ［数］［天］极坐标

pole /pəul/ n. 杆；极点；电极 vt. 用竿支撑

policy /'pɒləsi/ n. 政策，方针

polish /'pɒlɪʃ/ v. 抛光，擦亮 n. 磨光，擦亮；打磨光滑的面

pollute /pə'lu:t/ vt. 污染；玷污；败坏

pollution /pə'lu:ʃ(ə)n/ n. 污染，污染物

port /pɔ:t/ n. （计算机的）端口；左舷；舱门 vt. 持（枪）

practical /'præktɪk(ə)l/ adj. 实际的；实用性的

precision /prɪ'sɪʒ(ə)n/ n. 精度，［数］精密度；精确 adj. 精密的，精确的

presence /'prez(ə)ns/ n. 存在；出席；参加；风度；仪态

pressure /'preʃə/ n. 压力；压迫，［物］压强 vt. 迫使；密封；使……增压

preventive /prɪ'ventɪv/ adj. 预防性的，防备的 n. 预防药，预防疗法

price /praɪs/ n. 价格；价值；代价 vt. 给……定价；问……的价格

primarily /'praɪm(ə)rɪlɪ;praɪ'mer-/ adv. 首先；主要地，根本上

principle /'prɪnsɪp(ə)l/ n. 原理，原则；主义；本质；根源

print /prɪnt/ n. 印刷业；印刷字体 vt. 印刷；打印 vi. 印刷；出版

procedure /prə'si:dʒə/ n. 程序，手续；步骤

professional /prə'feʃ(ə)n(ə)l/ n. 专业人员；职业运动员 adj. 专业的；职业性的

programmable /ˌprəu'græməbl/ adj. ［计］可编程的；可设计的

programmer /'prəugræmə/ n. ［自］［计］程序设计员

programming /'prəugræmɪŋ/ n. 设计，规划；编制程序，［计］程序编制

projection /prə'dʒekʃ(ə)n/ n. 投射；规划；突出；发射；推测

prolonged /prə'lɒŋd/ adj. 延长的；拖延的；持续很久的

provision /prə'vɪʒ(ə)n/ n. 规定；条款；准备 vt. 供给…食物及必需品

pulse /pʌls/ n. ［电子］脉冲；脉搏 vt. 使跳动 vi. 跳动，脉跳

purpose /'pɜ:pəs/ n. 目的；用途；意志 vt. 决心；企图；打算

Industrial Robot

Q

quality /'kwɒlətɪ/ n. 质量，[统计] 品质；特性；才能 adj. 优质的；高品质的

query /'kwɪərɪ/ n. 疑问；疑问号；[计] 查询 vi. 询问 vt. 询问

R

realistic /rɪə'lɪstɪk/ adj. 现实的；现实主义的；逼真的；实在论的

realize /'rɪəlaɪz/ vt. 实现；认识到；了解；将某物卖得

real-time /ˌrɪəl'taɪm/ adj. 实时的；接到指示立即执行的

reassembly /ˌriː ə'semblɪ/ n. 重新聚集；重新组装；再装配

recall /rɪ'kɔːl/ n. 召回；回忆；撤销 vt. 召回；回想起，记起；取消

reciprocate /rɪ'sɪprəkeɪt/ vt. 报答；互换 vi. 往复运动；互换

rectangular /rek'tæŋgjulə/ adj. 矩形的；成直角的

reducer /rɪ'djuːsə/ n. [助剂] 还原剂；减径管

reference /'ref(ə)r(ə)ns/ n. 参考，参照 vt. 引用 vi. 引用

reflect /rɪ'flekt/ vt. 反映；反射；表达；显示 vi. 反射，映现；深思

reform /rɪ'fɔːm/ v. 改革；重组 n. 改革，改良；改正 adj. 改革的

refresh /rɪ'freʃ/ vt. 更新；使……恢复 vi. 恢复精神

register /'redʒɪstə/ n. 登记表；声区；套准 v. 登记

reinforce /riːɪn'fɔːs/ n. 加强；加固材料 vt. 加强，加固 vi. 求援；得到增援

relatively /'relətɪvlɪ/ adv. 相当地；相对地，比较地

reliable /rɪ'laɪəb(ə)l/ n. 可靠的人 adj. 可靠的；可信赖的

remain /rɪ'meɪn/ n. 遗迹；剩余物，残骸 vi. 保持；依然；留下；残存

remote /rɪ'məut/ n. 远程 adj. 遥远的；偏僻的；疏远的

repairer /ri'pɛərə/ n. 修理者；修补者

repeatable /ri'piːtəbl/ adj. 可重复的；可复验的

repetition /repɪ'tɪʃ(ə)n/ n. 重复；背诵；副本

reprogram /riː'prəugræm/ v. 为……重编程序

rescue /'reskjuː/ n. 营救，解救，援救 v. 营救，援救；

resistance /rɪ'zɪst(ə)ns/ n. 阻力；电阻；抵抗；反抗；抵抗力

resistive /rɪ'zɪstɪv/ adj. 有抵抗力的；抗……的，耐……的；电阻的

resolution /rezə'luːʃ(ə)n/ n. [物] 分辨率；决议；解决；决心

response /rɪ'spɒns/ n. 响应；反应；回答

reverse /rɪ'vɜːs/ v. 颠倒；反转 n. 逆向；相反 adj. 相反的；颠倒的

revival /rɪ'vaɪvl/ n. 复兴；复活；苏醒；恢复精神；再生效

revolution /revə'luːʃ(ə)n/ n. 革命；旋转；运行；循环

rigidity /rɪ'dʒɪdətɪ/ n. [物] 硬度，[力] 刚性；严格，刻板；僵化；坚硬

rivet /'rɪvɪt/ n. 铆钉 vt. 铆接；固定；集中于

robot /'rəubɒt/ n. 机器人；机械般工作的人

roller /'rəulə/ n. [机] 滚筒；[机] 滚轴；辊子；滚转机

rotary /'rəut(ə)rɪ/ n. 旋转式机器 adj. 旋转的，转动的；轮流的

rotation /rə(ʊ)'teɪʃ(ə)n/ n. 旋转，循环，轮流

round /raund/ adj. 圆的；弧形的 n. 阶段；轮次 prep. 围绕；绕过

rout /raut/ v. 彻底击败，打垮；刻纹 n. 溃败；溃退

routine /ruː'tiːn/ n. 常规，惯例 adj. 常规的，例行的 v. 按惯例安排

S

safety /'seɪftɪ/ n. 安全；保险；安全设备

salary /'sælərɪ/ v. 给……薪金 n. 薪水，工资

sample /'sɑːmp(ə)l/ v. 品尝，体验 adj. 样品的 n. 样品

sand /sænd/ n. 沙；沙地 vt. 擦平或磨光某物

satisfy /'sætɪsfaɪ/ vi. 令人满意；令人满足 vt. 满足；说服；使满意

scan /skæn/ n. 扫描；浏览 vt. 扫描；浏览 vi. 扫描；扫掠

schedule /'skedʒul/ v. 安排，预定 n. 计划（表）；时间表

schematic /skiː'mætɪk; skɪ-/ n. 原理图；图解视图 adj. 图解的；概要的

scientific /saɪən'tɪfɪk/ adj. 科学的，系统的

scope /skəup/ n. 范围；余地；视野；眼界 vt. 审视

segment /'segm(ə)nt/ n. 段，部分 v. 分割

seize /siːz/ vt. 抓住；夺取；逮捕 vi. 抓住；利用

sense /sens/ n. 感觉，功能；理智 vt. 感觉到；检测

sensitivity /sensɪ'tɪvɪtɪ/ n. 敏感；敏感性；过敏

sequence /'siːkw(ə)ns/ n. [数] [计] 序列；顺序；续发事件 vt. 按顺序排好

series /'sɪəriːz; -rɪz/ n. 系列，连续；[电] 串联；级数；丛书

serve /sɜːv/ n. 发球 vi. 服役，招待，侍候 vt. 招待，供应；为…服务

servo motor n. 伺服马达；伺服电动机

shaft /ʃɑːft/ n. 竖井；通风井杆，柄

shallow /'ʃæləʊ/ n. [地理] 浅滩 adj. 浅的；肤浅的 vt. 使变浅 vi. 变浅

shape /ʃeɪp/ n. 形状；模型 vt. 形成；塑造 vi. 形成；成形；成长

shortcoming /'ʃɔːtkʌmɪŋ/ n. 缺点；短处

shoulder /'ʃəʊldə/ n. 肩，肩膀；肩部 vi. 用肩顶 vt. 肩负，承担

shutdown /'ʃʌtdaun/ n. 关机；停工；关门；停播

shutter /'ʃʌtə/ n. 快门；百叶窗；遮板 vt. 以百叶窗遮蔽

side /saɪd/ n. 方面；侧面；旁边 adj. 旁的，侧的 vt. 同意，支持

sight /saɪt/ n. 视力；景象 adj. 即席的 vt. 看见 vi. 瞄准；观看

signal /'sɪgn(ə)l/ n. 信号；暗号 adj. 显著的 vt. 标志；表示 vi. 发信号

size /saɪz/ n. 大小；尺寸 adj. 一定尺寸的 vt. 依大小排列 vi. 可比拟

slider /'slaɪdə/ n. 滑动器，滑竿；浮动块，滑动块

sliding /'slaɪdɪŋ/ n. 滑；移动 adj. 变化的；滑行的 v. 滑动；使滑行

Industrial Robot

slipping /'slɪpɪŋ/ n. 滑动 v. 滑动 adj. 渐渐松弛的

slowdown /'sləʊdaʊn/ n. 减速；怠工；降低速度

smell /smel/ n. 气味，嗅觉 v. 嗅，闻；察觉到

smooth /smuːð/ n. 平滑部分 adj. 平稳的 vt. 使光滑 adv. 光滑地；平稳地

software /'sɒf(t)weə/ n. 软件

solution /sə'luːʃ(ə)n/ n. 解决方案；溶液；溶解；解答

sophisticate /sə'fɪstɪkeɪt/ adj. 老于世故的 v. 弄复杂；曲解

space /speɪs/ n. 空间；太空；距离 vt. 隔开 vi. 留间隔

spectral /'spektr(ə)l/ adj. ［光］光谱的；幽灵的；鬼怪的

speed /spiːd/ v. 快速运动；加速；（使）繁荣；n. 速度；进度；迅速

spherical /'sferɪk(ə)l/ adj. 球形的，球面的；天体的

spot /spɒt/ n. 地点；斑点 adj. 现场的 vt. 认出；弄脏

spray /spreɪ/ n. 喷雾，喷雾剂；喷雾器；水沫 vt. 喷射 vi. 喷

sprocket /'sprɒkɪt/ n. 链轮齿；扣链齿轮

stack /stæk/ n. （整齐的）一堆；垛，堆 v. （使）成叠地放在

state /steɪt/ n. 国家；州；情形 adj. 国家的；正式的 vt. 规定；声明；陈述

static /'stætɪk/ n. 静电；静电干扰 adj. 静态的；静电的；静力的

stationary /'steɪʃ(ə)n(ə)rɪ/ n. 不动的人 adj. 固定的；静止的；定居的

status /'steɪtəs/ n. 地位；状态；情形；重要身份

steam /stiːm/ n. 蒸汽；蒸汽动力 vi. 冒水汽 adj. 蒸汽的

stepper /'stepə/ n. 分档器

stick /stɪk/ n. 棍；手杖 vi. 坚持；伸出；粘住 vt. 刺，戳；粘贴

storage /'stɔːrɪdʒ/ n. 存储；仓库；贮藏所

store /stɔː/ n. 商店；储备，贮藏；仓库 vt. 贮藏，储存

stretch /stretʃ/ v. 伸展；拉紧 adj. 弹性的，可拉伸的 n. 舒展；伸张

strict /strɪkt/ adj. 严格的；绝对的；精确的；详细的

stroke /strəʊk/ n. 冲程；笔画；打击 vt. 抚摸；敲击 vi. 击球

structure /'strʌktʃə/ n. 结构；构造；建筑物 vt. 组织；构成；建造

stud /stʌd/ n. 种马；大头钉；饰纽；壁骨 vt. 散布；用许多饰纽等装饰

subdivide /sʌbdɪ'vaɪd/ vt. 把……再分，把……细分 vi. 细分，再分

substation /'sʌbsteɪʃ(ə)n/ n. 分局；变电所；分所；分台

subsystem /'sʌbsɪstəm/ n. 子系统；次要系统

sufficient /sə'fɪʃ(ə)nt/ adj. 足够的；充分的

supervise /'suːpəvaɪz;'sjuː-/ vt. 监督，管理；指导 vi. 监督，管理；指导

supervision /ˌsuːpə'vɪʒn;ˌsjuː-/ n. 监督，管理

supplier /sə'plaɪə/ n. 供应厂商，供应国；供应者

surgeon /'sɜːdʒ(ə)n/ n. 外科医生

surgical /'sɜːdʒɪk(ə)l/ n. 外科手术；外科病房 adj. 外科的；手术上的

surrounding /sə'raʊndɪŋ/ n. 环境，周围的事物 adj. 周围的，附近的

Sweden /'swidən/ *n.* 瑞典

switch /switʃ/ *n.* 开关；转变 *v.* 改变（立场、方向等）；替换；调换

synchronous /'sıŋkrənəs/ *adj.* 同步的；同时的

synchronous belts *n.* 同步带

syntax /'sıntæks/ *n.* 语法；句法；有秩序的排列

systematic /sıstə'mætık/ *adj.* 系统的；体系的；[图情] 分类的；一贯的

T

tachometer /tæ'kɒmıtə/ *n.* 转速计，转速表

tactile /'tæktaıl/ *adj.* [生理] 触觉的，有触觉的；能触知的

talent /'tælənt/ *n.* 才能；天才；天资

tank /tæŋk/ *n.* 坦克；水槽；池塘 *vt.* 把…贮放在柜内；打败 *vi.* 乘坦克行进

target /'tɑːgıt/ *n.* 目标，指标；靶子 *v.* 把…作为目标；面向

taste /teıst/ *n.* 味道；品味；审美 *vt.* 尝；体验 *vi.* 尝起来

technical /'teknık(ə)l/ *adj.* 工艺的，科技的；技术上的；专门的

technician /tek'nıʃ(ə)n/ *n.* 技师，技术员；技巧纯熟的人

technology /tek'nɒlədʒı/ *n.* 技术；工艺；术语

tedious /'tiːdıəs/ *adj.* 沉闷的；冗长乏味的

temperature /'temprətʃə(r)/ *n.* 温度

tension /'tenʃ(ə)n/ *n.* 张力，拉力 *vt.* 使紧张；使拉紧

terrain /tə'reın/ *n.* [地理] 地形，地势；领域；地带

text /tekst/ *n.* [计] 文本；课文；主题 *vt.* 发短信

thermistor /θɜː,mıstə/ *n.* [电子] 热敏电阻；电热调节器

threat /θret/ *n.* 威胁，恐吓；凶兆

tidy /'taıdı/ *adj.* 整齐的 *vi.* 整理；收拾 *vt.* 整理；收拾

time /taım/ *n.* 时间；时代 *adj.* 定时的 *vt.* 计时

title /'taıt(ə)l/ *n.* 标题；头衔；权利 *adj.* 标题的 *vt.* 加标题于

touch /tʌtʃ/ *v.* 接触；轻按；轻弹；相互接触 *n.* 触碰；轻按；触觉

tough /tʌf/ *adj.* 艰苦的，困难的 *vt.* 坚持；忍耐 *adv.* 强硬地，顽强地

toxic /'tɒksık/ *adj.* 有毒的；中毒的

trajectory /'trædʒıkt(ə)rı/ *n.* [物] 轨道，轨线；[航] [军] 弹道

transducer /trænz'djuːsə;-ns-/ *n.* [自] 传感器，[电子] 变换器

transfer /træns'fɜː/ *v.* 转让；转接；移交 *n.* （地点的）转移

transmission /trænz'mıʃ(ə)n/ *n.* 传动装置，[机] 变速器；传递；传送

trigger /'trıgə/ *n.* 扳机；起因；触发器 *v.* 触发；开动（装置）

troubleshooter /'trʌbl,ʃuːtə/ *n.* 解决纠纷者；故障检修工

tune /tjuːn/ *n.* 曲调；和谐 *vt.* 调整；使一致 *vi.* [电子] [通信] 调谐；协调

twisting /'twıstıŋ/ *n.* 快速扭转，缠绕 *v.* 使弯曲 *adj.* 曲折的，缠绕的

U

ultra　/'ʌltrə/ *adj.* 极端的，偏激的 *n.* 过激分子，极端主义者 *adv.* 很，非常

unique　/juːˈniːk/ *n.* 独一无二的 *adj.* 独特的，稀罕的；[数] 唯一的

universal　/juːnɪˈvɜːs(ə)l/ *n.* 一般概念；普遍性 *adj.* 普遍的；通用的；全体的

unload　/ʌnˈləud/ *vt.* 卸；摆脱…之负担；倾销 *vi.* 卸货；退子弹

V

variety　/vəˈraɪətɪ/ *n.* 多样；种类；杂耍；变化，多样化

V-belts　*n.* 三角 V 带，V 形带

velocity　/vəˈlɒsəti/ *n.* [力] 速率；迅速；周转率

vertical　/ˈvɜːtɪk(ə)l/ *n.* 垂直线，垂直位置 *adj.* 垂直的，直立的

vibration　/vaɪˈbreɪʃ(ə)n/ *n.* 振动；犹豫

vice versa　/ˌvaisiˈveːsə/ *adv.* 反之亦然

video　/ˈvɪdɪəu/ *n.* [电子] 视频；录像，录像机 *adj.* 视频的 *v.* 录制

vision　/ˈvɪʒ(ə)n/ *n.* 视力；美景；眼力；想象力 *vt.* 想象；显现；梦见

visual　/ˈvɪʒjuəl;-zj-/ *adj.* 视觉的，视力的

voltage　/ˈvəultɪdʒ;ˈvɒltɪdʒ/ *n.* [电] 电压

W

wafer　/ˈweɪfə/ *n.* 圆片，晶片；薄片，干胶片；薄饼 *vt.* 用干胶片封

wage　/weɪdʒ/ *v.* 进行，发动，开展 *n.* 工资；报酬；代价

warehouse　/ˈweəhaus/ *n.* 仓库；货栈；大商店 *vt.* 储入仓库

wear　/weə/ *v.* 穿，戴；耐用 *n.* 衣物；磨损；耐久性

weld　/weld/ *n.* 焊接；焊接点 *vt.* 焊接；使结合；使成整体 *vi.* 焊牢

wheel　/wiːl/ *n.* 车轮；转动 *vt.* 转动；使变换方向 *vi.* 旋转

wire　/waɪə/ *n.* 电线；金属丝；电报 *vt.* 拍电报；给…装电线 *vi.* 打电报

workflow　/ˈwəːk,fləu/ *n.* 工作流，工作流程

working envelope　*n.* 工作包络面

workstation　*n.* 工作站

wrist　/rɪst/ *n.* 手腕；腕关节 *vt.* 用腕力移动

References

[1] Miller M R, Miller R. Robots and Robotics Principles, Systems, and Industrial Applications [M]. New York: McGraw-Hill Education, 2017.

[2] Todd D J. Fundamentals of Robot Technology: An Introduction to Industrial Robots [M], New York: Wiley, 1986.

[3] Roshid M M. Introduction to Industrial Robot: A briefly description of Industrial Robot [M], Saar brücken: VDM Verlag Dr. Müller, 2011.

[4] Appleton E, Williams D J. Industrial Robot Applications [M]. Berlin: Springer, 2011.